BUILDING THE
BEAM ENGINE MARY

BUILDING THE BEAM ENGINE MARY
Tubal Cain

Model & Allied Publications, Argus Books Limited

Model & Allied Publications
Argus Books Ltd
14 St. James Road
Watford, Herts, England

© Argus Books Ltd, 1981
© Tubal Cain, 1981

All rights reserved. No part of this book may be reproduced without the written permission of the publishers.

ISBN 0 85242 754 9

The text of this book is based on articles from the *Model Engineer*.

Typeset by Computer Photoset, Birmingham
Printed by Pindar Print Ltd, Scarborough

The first 'MARY', as built by the Author. (Unpainted when this photograph was taken).

'MARY'
A Four-Column Beam Engine

The Beam Engine is the earliest form of steam engine, having been introduced by Thomas Newcomen in 1712. The arrangement followed the existing practice of water-wheel pumping engine. This was highly developed at the beginning of the 18th Century and it was quite logical that the engineers of the day should regard the floor as the 'proper' place to site both cylinder and crankshaft. From 1800 onwards the beam engine developed into a very reliable prime mover for all purposes—even, in America, as a marine engine—and many of the examples were both magnificent and elegant. It is not surprising that such engines as still exist are always strong candidates for preservation. Their slow speed enables all the working mechanism to be clearly followed by the eye and the best of them have a grace and dignity not to be found in any other type of engine. 'MARY' is an example of a relatively small one—'Twenty Horses Power' would have been the catalogue description—and the scale is such that no difficulties should be encountered in manufacture. She is representative of mid 19th Century practice, which many regard as the 'Age of Elegance' in steam engine design.

By 1850 the majority of steam engines built were of the horizontal type, but there were still areas in which the slow-speed, solidly built beam type was preferred. Condensing type engines were employed still in the large textile mills; non-rotative pumping engines survived in waterworks for another 50 or 60 years, as well as for ironworks blowing engines, though these latter usually were fitted with cranks and flywheel. The non-condensing rotating engine held sway in the sugar mills of the West Indies. The absence of a condenser was not due to any neglect of thermal efficiency on the part of the users; far from it. As early as 1803, Trevethick had called attention to the use of the exhaust steam from a 'high-pressure' engine for heating purposes and did, in fact, supply such engines for sugar mills, where the process involved the heating of the raw sugar syrup. This arrangement was, in fact, more efficient than the application of a condenser to the engine, with a separate source of heat for the vats. Such engines were made until quite late in the 19th century, by Mirrlees & Tait, McOnie of Glasgow, Fawcett Preston & Co and Forresters of Liverpool. The latter firm, indeed, exhibited a steam driven *vacuum* pump for use on sugar refiners at the great Exhibition of 1862, the combined engine and pump being a beam engine.

Some difficulty was experienced in selecting a suitable prototype, chiefly because the most attractive LOOKING engines are often very large, whilst some of the smaller ones had the decoration carried to the extreme of embellishment. Some of the latter were very interesting indeed, but I could forsee a howl of protest from the foundry were the pattern-maker asked to carve the main-bearings in the shape of a pair of dolphins!

There was the further point that I

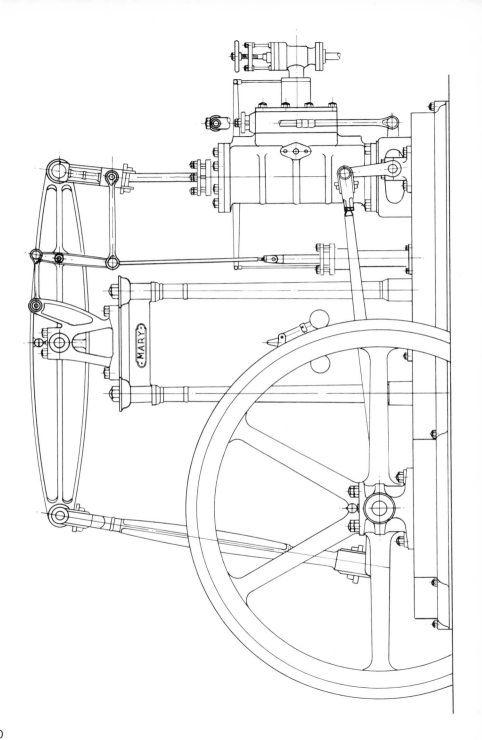

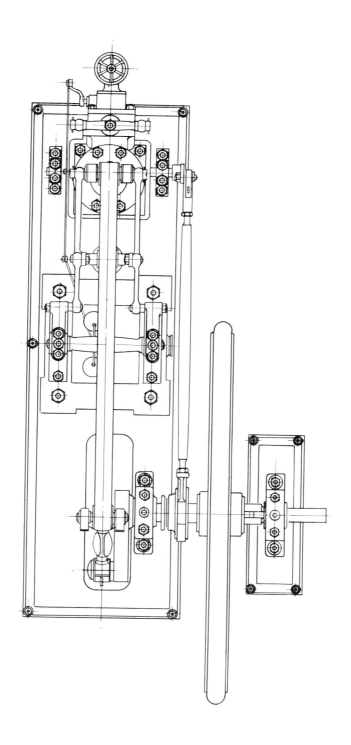

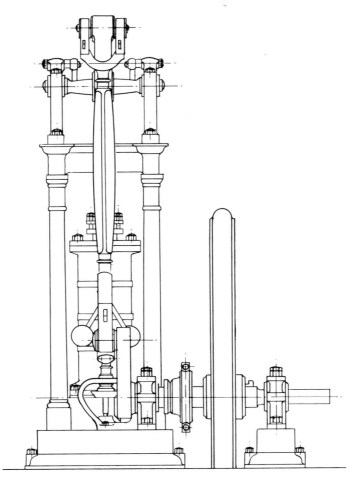

General Arrangement, from the design drawing. (See also previous pages).

wanted, if possible, to include some unusual technical point of design, and many of the prototypes of which information was available were sadly 'run of the mill' in this respect. In the event, the engine is a combination of three engines; a sugar-mill engine by Peel Williams, a contemporary model of a welsh tin-plate mill engine which I have, makers unknown, and a very small engraving of another engine, again of unknown make, but which exhibited the unusual form of parallel motion shown on the model. Some such device, by the way, is almost essential if only four columns are to be used, and an 'outrigger' to carry the standard 'Watt' parallelogram motion link is to be avoided. The only reason for *six* columns, as normally used, is for the support of the necessary entablature to carry this motion, though the mill-wrights often found this an advantage if ever the crank had to be lifted—tackle could be supported on the entablature.

I have taken a few liberties in the design—this, by the way, was done full size, and then reduced to model proportions. Almost all of these are due to the difficulty of scaling small details. The parallel motion links, for example, should all have split and cottered bearings. The valve-chest should be a closed box bolted to the cylinder-face; this would have led not only to difficulties of timing the valves through a 5/32" hole, but also to the use of 14BA studs! The governor pivots should all have pins with collars and taper-pins, but the latter would have been only .015 in. diameter if used,—and the pivots themselves about $\frac{3}{64}$ in. diameter. The general appearance of the engine is, however, true enough to engines of the 1850's to satisfy all but the most meticulous.

The general arrangement drawings are shown on pages 10, 11 and 12. This, by the way, is a tracing of the 'design' drawing, and a few of the details may differ slightly when they come to be described. (In particular, the governor is taller, as one cannot scale the height of a Watt governor). The parallel motion works on the same *principle* as the 'normal', but the necessary reverse arc of motion is obtained from a short link running backwards to a pivot on the beam support hearing *above* the beam, in place for the normal link terminating on the cylinder centreline *below* the beam. This is equally effective; none of these parallel motions gives a truly straight-line travel to the crosshead, and allowance has to be made for this in designing the gland. The feed pump drive is from the inner end of this motion, as usual. The pump, by the way, is a dummy, but those wishing to make a working pump. will find sufficient room for the valve-box inside the bedplate.

The steam distribution is by a normal slide-valve—somewhat shorter than was used on the prototype—driven through the usual pattern of rocker-shaft from a single eccentric. (Some of the sugar-mill engines had double-eccentric 'expansion' valve gears fitted, but it is doubtful if they served much useful purpose as such engines usually ran on full power). Some readers may comment on the absence of a 'gab' on the eccentric shaft, often seen on models, to enable the engine to be worked by the hand-gear. By mid-nineteenth century, however, engine driving was an established art, and this device was no longer needed on rotative engines. The Governor is a standard 'Watt' conical pendulum, but speeded up a little from the crank, and with links designed to give a slight hope of the governor actually controlling the engine. This point will be dealt with in more detail when the governor construction is described. The correct form of cotter-and-gib adjustment is provided on the connecting rod, parallel motion main links, and (though with no adjustment) on the eccentric rod.

The only ornament I have 'designed' is that on the base and entablature; this is as near correct from the real engine as I could measure. That on the columns is also taken from the prototype, but there is no reason why other forms of corinthian decoration cannot be used— these profiles are machined onto the columns. The arrangement drawings show plain nuts on the entablature, but the detail suggests nuts of 'acorn' shape; again, the builder can express his individuality if he wishes, as also on the vertical valve-operating links and the parallel motion too. The only rule to be observed is that of 'restraint'. The entablature is in what is known as a 'severe classical' style and would not be associated with ornate embellishment—such as column nuts in the shape of pineapples. (I have a drawing of an engine with these, the beam bearings being supported on what appear to be two sprays of rhododendron leaves!)

As to supplies, Messrs Reeves of Birmingham can supply the castings needed; most of these are in gunmetal. Bar material needed is conventional, mainly brass or mild steel with rustless steel for the piston and valve-rods. (If you wish to use rustless throughout this is in order, of course) I have designed all glands and the piston for silicone rubber 'O' rings. Reeves can supply the gears for the Governor but these will

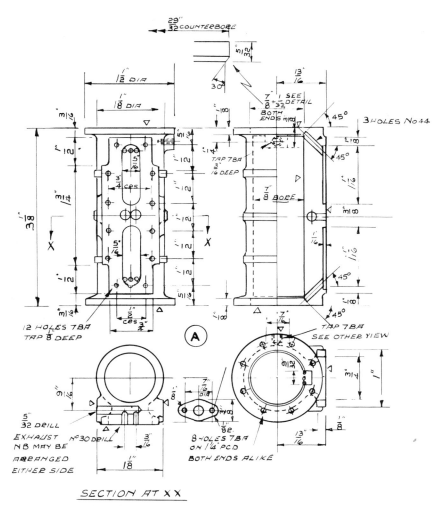

Fig 1-Details of cylinder and top cover.

need modifying in the boss—not a difficult job. There are on the engine a number of parts held with pinned collars, and these present two problems—they are small, and few of the normal taper-pin broaches fit even standard pins anyway.

These small pins, and the associated taper broaches, can be had from Messrs A. G. Thomas, Tompion House, Heaton Road, Bradford, or from H. S. Walsh & Co., 12 Clerkenwell Road, London, EC1. These little broaches are really Horological tools, and I have found them invaluable for many jobs around the workshop.

An engine like this needs a protective case, as otherwise an accumulation of oily dust will soon choke up the more delicate moving parts, as well as being unsightly. Messrs Visijar, Ltd, of Corringham Road, Gainsborough, Lincs DN21 1QB, can provide this. My own is $16\frac{1}{2}$ in. long by $8\frac{1}{2}$ in. wide and $14\frac{1}{2}$ in. tall, and if you make the underbase to suit with holes for inlet and exhaust pipes you

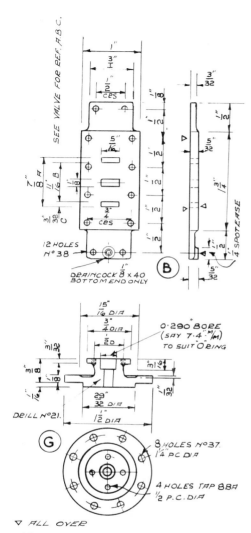

a 'design office' drawing! First, you will find that most castings have one or both of two symbols on some surfaces. A little triangle, or a letter 'f'. The triangle is an instruction to the pattern-maker to provide enough metal for machining; the 'f' means enough metal to allow cleaning up with a file. This doesn't mean you have got to *machine* to the triangle—use a file if you like; on the real engine many would have been done with hammer and chisel. Likewise you can machine instead of filing if you want to. The marks are really for the patternmaker—all *you* have to do is to bring the part to the right dimension! Secondly, there may be some dimensions shown which have no relevance to *your* work. These again are for the pattern-maker. As far as I can I have deleted all these from the final drawings, but if you *do* find that a little fillet is marked '$\frac{1}{8}$ in. rad' it doesn't mean you have to fiddle away with a ball-ended end-mill to get to that figure exactly!

All parts have an identification letter, and these refer to a schedule of parts which will appear on the full drawings. The repetition of some part letters for different items is due to there being four sheets of detail drawings. These schedules (see page 80) will list all the parts needed—I hope. (I do know that a $\frac{3}{32}$ in. hard steel ball in the governor got left off, but I think all the rest is there.) You will be able to use these schedules to order up your nuts and bolts—and studs if you are not going to make them. (Studs can be had from Stuart-Turners, ready machined; a great saving of time.)

Finally, please note that there is no reason at all why you should not depart from the dimensions given provided you keep a sense of proportion about it. For example, the crank is carried in two bearings of different diameters; there is no reason, other than the fact that the crank on the prototype was like that, why you shouldn't make them both the large diameter. The engine will not collapse if you use 8BA studs instead of 7BA, and if you have plenty of one

will be able to run the engine on compressed air (not steam!) whilst still under cover.

Now for the drawings. As well as appearing in these pages as the description proceeds, they will be supplied with the castings, as a complete set. I must say a word about these, as they depart a little from the type of drawing usually supplied for model making. Old habits die hard, and I find it takes me nearly twice as long to make other than

size and not the other, this would be sensible—again, within reason. A cotter strap shown $\frac{1}{16}$ in. thick can just as easily be 1.5 mm if that's what you have in the stores. The dimensions you *mustn't* alter are those which affect the dynamics of the engine, or which would alter the valve events or the clearance at the end of the cylinder. Especially you must take care about the lengths of the links in the parallel motion; alter the widths, or the sizes of the pivotpins if you like, but keep to the design lengths or it won't work! Now to the construction.

General

The first job I always do is to go over all castings trimming them up with a file—saves time later and enables one to get familiar with them. Next, I usually set up the 4-jaw and go through all parts which have chucking pieces (cylinder covers, glands, etc.) and machine these chucking pieces true—sometimes also facing the back face of the casting as well. This enables a lot of subsequent work to be done in the three-jaw, and it can be time-consuming having to make frequent chuck changes, as well as being bad for the mandrel nose. Finally, I usually spray all castings with a coat of cellulose primer from an aerosol 'automobile spray' both to prevent rust on iron, to prevent oil getting into the pores of gunmetal, and to serve as a base for scribed lines. Which castings you start to machine is up to you, but I usually plan to handle all gunmetal first, and then the iron, finally going on to steel bar work, if only to save tool-changing.

Cylinder Set (Fig. 1)

There are 'n' ways of boring a cylinder, where 'n' is the number of model engineers. This is the way I did this one, which is rather long for its bore. Knock a lead or close-grained hard-wood plug into the ends of the cylinder and set up between centres, *very* lightly gripped.

Adjust the position till the *outside* of the cylinder barrel runs as true as you can get. Very carefully screw in the tailstock centre; it doesn't matter if the casting moves a few thou., but try again if the shift is more than $\frac{1}{64}$ in. Deepen the centres with a centre-drill by hand if you like. Apply a carrier, and machine the flanges true, to within, say .010 in. of finished diameter and within about the same amount or a little more of the correct face-to-face length. Take lightish cuts, to avoid the risk of the centres working in their holes. Remove from the machine and knock out the plug from the *top* end of the cylinder. (This end has a little boss to carry the governor linkwork.) Set this end in the 4-jaw, with 10 or 15 thou. shimbrass under the jaws, and reset till *both* flanges are running true—the outer (bottom) one dead true, that in the chuck as near as you can get. The centre remaining in the plug that's still there will help in the first stages. Don't grip too hard in the chuck, or you will distort the casting.

Now set up your fixed steady on the outer flange, using cigarette paper to check that you are touching without moving the casting. As soon as you are satisfied, see that the steady nuts are tight, and remove—gently—the plug in the bore. Set up the stiffest boring bar you have in the toolholder and rough out the bore to within about $\frac{1}{64}$ in. of size. Don't let the work get too hot, or it will expand and distort in the chuck and steady. Go and have a smoke—or something—and then take out the tool and hone it to a really fine finish. Aim at a shape which gives about 30° lead angle in plan, with no more than $\frac{1}{64}$ in. flat or large radius on the face. Set on a feed of about 8 thou. per rev., and finish bore to size. Use a piece of $\frac{7}{8}$ in. PGMS or silver steel as a gauge—if you haven't any, you should have made a little plug gauge, with diameters 0.873 and 0.875, before you started to bore; but I used Silver steel. Get it as nearly right as you can, as 'O' rings do need a closer

Fig 2-Boring the cylinder. Note the stiff bar used.

tolerance to the bore than normal packing. Finally, engage your leadscrew handwheel (disconnect the auto-feed first!) and apply a cut of 15 thou with the cross-slide index, advance the tool $\frac{5}{32}$ in. into the bore to cut the counterbore at the cylinder end; then, retract the tool, feed right down to the other end, reset the cut, and counterbore the far end. Make quite sure that the tool has the correct plan angle to cut both ways, of course. Take a light skim over the face of the flange. Fig. 2 shows boring in progress.

Remove from the 4-jaw and either on a mandrel, or gripping lightly internally with the 3-jaw and using a large centre, bring the flanges to the correct diameter and face to face distance, removing any grip-marks from the chuck as you do so. Remove from the machine and take off any burrs and sharp edges.

We must now machine the flat portface, Fig. 3. Find a piece of $\frac{7}{8}$ in. bar which is free from burrs and which will slide smoothly into the bore. Set this up on Vee-blocks and packing till the bore is just at centre-height and the face vertical. Set a couple of clamps *on the casting*, and tighten down; this will hold both casting and bar secure, but to be safe set a little toolmakers jack—or a short $\frac{3}{8}$ in. bolt with nut part on to serve as a jack—under the front edge of the port-face, but not projecting. Make sure all is square across the lathe—measure from the bar to the chuck face. Set up a flycutter as shown in fig 3, or a large

Fig 3-Machining the portface, using a simple flycutter.

endmill if you have one, and machine the face. Substitute a $\frac{5}{16}$ in. endmill and cut the two grooves which will form the steam passages. The depth of these is not critical. Finally, whilst still set up true and to a known centreheight, use your scribing block or height gauge to mark out the centres for the two exhaust holes ($\frac{3}{16}$ in. apart).

Remove from the machine, and file up the face of the chosen exhaust flange—you can use either. If you wish, chew off the other one entirely. Set up on a surface plate, mark out for the $\frac{5}{32}$ in. exhaust hole and drill these. Mark out for and drill the two No. 30 holes to break into the $\frac{5}{32}$ in. one, and no further. Take care over these three holes, as they are close to the cylinder bore. Drill for the exhaust flange studholes, but NOT for that for the governor peg at this stage—just file (or mill) that flange true and leave till later. Finally, file or mill the 45° bevels at the cylinder ends, mark out for and drill the no. 44 holes for the steam passages. You can do this quite easily by resting the cylinder in the angle of a large vee-block. The rest of the cylinder holes must wait till you have machined the mating parts, to use as drill-jigs.

You will note that I have asked you to bore the hole and face the flange at one setting, with the base of the cylinder outwards. This is to ensure that the cylinder sits truly square on the bed-plate. This is important, and I shall be having a word about alignment later. For the same reason, the bottom cover thickness must be carefully watched when you come to it.

Top Cover, Part G

In this case you should have brought the spigot and the flange almost to size when you skimmed the chucking piece, but if not, do this, and re-skim before gripping by the chucking piece to turn the top. This may need a bit of wangling to get into the recess, but the *shape* is more important in here than exact sizes. Leave 5 thou. or so on the flange. Centre and drill the No. 21 hole, and follow with $\frac{9}{32}$ in., not quite to depth. With a little boring tool open out the bore to size—I found that the *shank* of a letter M drill was just right as a gauge. Mark out for the boltholes whilst in the chuck. Set your scribing block at centre-height and scribe right across; set this line vertical, and repeat. Now put the square in the Vee of a vee-block, to get it at 45°, set one of the lines to this, and scribe again. Repeat for the final line. Set the block to centre-height *less* $\frac{5}{8}$ in. to get the $1\frac{1}{4}$ in. PCD and mark for each hole in turn. Leave those in the gland till you have the latter as a template.

Reverse in the chuck and skim the spigot and bottom face, taking the usual precautions to get it true, and also trim the O.D. to fit that on the cylinder. In the case of this spigot, it should be left a few thou. *Under*size—say .003 in.—to allow for adjustment on assembly.

Port Cover, Part B

This is not an easy one to hold. If you have a shellac chuck you can try this method, but I think the grip will be insufficient on the discontinuous cut. I did not risk it, but used solder instead, fig 4. File the front (outer) face reasonable flat, and solder this to your faceplate or a spare backplate (if using one of these, skim it first; it may not be true). With a very sharp tool skim the face till the $\frac{3}{4}$ in. wide ends are to thickness. Reverse and re-solder to the faceplate. NOTE—make sure the faceplate is quite cold before putting onto the lathe mandrel, or it may shrink on! Skim the main body to thickness as well as the little boss for the draincock.

Leave the piece on the plate to help

Fig 4-The port cover soldered to a spare chuck backplate.

hold it whilst marking out for the ports. This must be done reasonable carefully, though it IS possible to correct for any errors by altering the valve. If you have no endmill of the correct size, then you must use the old and well tried method of drilling holes rather smaller than the port width and filing with a needle file; not so tedious as it seems, and relatively easy in this case, as the file can go right through. (In fact, this method is probably quicker overall than setting up in a milling machine and tweedling out the ports!) I did mine in the lathe, setting up the piece (unsoldered from the backplate) in a vice on the vertical slide, and using the indexes on vertical slide and cross-slide to get the ports in the right position and the right width—no need to mark out at all except to establish the centrelines. The only precaution to be taken is to realise that an end-mill *won't* drill, it doesn't cut at the centre, so you must use a slot-drill or drill a hole—quite small—first. Don't attempt any cuts deeper than about $\frac{1}{4}$ of the diameter of the cutter, and run as fast as you can. (2500 rpm is about right for a $\frac{3}{32}$ in. cutter in gunmetal.)

Now file the outline more or less to suit the profile of the cylinder, but don't drill the fixing holes till you have made the steam chest to use as a template. You can, however, drill and tap for the drain cock at this stage, and if, when you offer it up, this hole is nearly 'blind', cut out a little nick in the end of the steam passage on the cylinder to make a waterway.

Assuming you have the steamchest drilled, use this to spot through for the holes in the port cover, and then that, in turn, to spot through for the holes in the cylinder. In all cases clamp the parts together, don't rely on 'hand holding'. Drill and tap the cylinder. Tin the face of the cylinder and at the same time reheat the cover—wipe both with a piece of clean rag on which a little powdered resin has been sprinkled, whilst still hot, to get a clean tinned surface. Apply a little NON ACID flux and a trace of 'solder paint', the latter only round the portways, and attach the two parts with screws which have been dipped in graphite, blacklead (allowed to dry) or failing all else blackened in a candle-flame. Heat the parts sufficient to melt the solder, and then wipe the outside, and allow to cool naturally—don't 'quench' it. Remove the screws. You can now countersink the holes in the two ends if you like—those at $\frac{1}{2}$ in. centres—and fit countersunk screws, the heads to be filled over when painting.

If you wish, there is no reason why you should not unite these two parts with a paper joint in the ordinary way. The only reason for using this construction is to avoid having to drill three tiny holes over an inch long at an angle down the cylinder. However, if you do use the solder method, the last job is to boil all in water to dispose of any stray flux. Finally, file and trim so that the combination looks like one piece.

Steam Chest, Part D, and Cover, Part E (Fig. 5)

There is very little work on these apart from filing or milling the surfaces, as the machining of the gland is the same process (though a different size) as that on the cylinder cover. When holding this part in the 4-jaw, don't forget to use soft packing as otherwise the material will mark—it is pretty soft. You can machine the whole issue in the lathe if you wish—just a question of preference, and which set-up you have in the lathe at the time. The only point worth mentioning is, perhaps, that the *inside* of the steam chest is best brought to size first, so that the drilling of the tailrod hole is facilitated by having a smooth surface presented to the drill. Don't mark out for the gland screw holes till that part is available as a template, the same applies to the steam flange holes—these will be marked out from the governor throttle casing. I marked out the fixing holes on the chest and used this as template for the cover and, as already noticed, on the port cover. Don't forget to mark mating parts to get them right way round.

Valve, Part C (Fig. 5)

This can be milled or filed—the latter is quicker and just as good. First, bring the face of the valve to flat, with the cavity the correct depth, and then file the back to the proper dimension. Now file the sides both to correct width and with the cavity central. Clean up the outside of the valve-rod boss, but leave a bit on the length at this stage. The only part I prefer to (and did) mill is the valve-rod slot; this is best done by gripping in the vertical slide vice and running a slot-drill or endmill down, but failing this you can drill a row of small holes and clean up to an easy sliding fit on a piece of $\frac{3}{32}$ in. silver steel, or a $\frac{3}{32}$ in. drill at a pinch. It is fairly important that this slot be true to the sides.

Now for the 'business faces'—the ends. First check the length of the cavity against the distance 'B' on the portface of *part* B. It must be corrected with care and a fine file till it is the same or *very* slightly (a thou. or so) less *and* square to the sides of the valve. Now decide which end is the 'top' and mark it. File the length of the valve till it is about $\frac{1}{64}$ in. oversize. The cavity central, and square to the sides, and put on the bevel shown on the drawing. Now check the top port, dimension C on the drawing ($\frac{3}{32}$ in.) and file that end of the valve till 'Z' is between 30 and 34 thou. larger. Similarly the bottom edge. The valve will now give the correct timing even if the ports are not quite right. Incidentally, if the cavity is too *long*, so that both ports will be open at the same time, there is little you can do about it except to ask for a new casting. Once the valve is the correct size, bring the working face dead flat, remove sharp edges all round *except* of the working face, and wrap it up and put it in a safe place. (Make a note on the drawing where it is, or you may forget!)

Bottom Cover, Part H (Fig. 5)

This needs little comment except to take light cuts with a sharp tool or it may distort. Mark out for the holes as for the top cover, and for the edges, taking care the holes are correctly placed in relation to the sides. When machining the back, make sure the thickness is within .002 in. as this affects alignment. You may now drill both covers, spot through onto the cylinder, and drill and tap the latter. Take care that the bottom cover side lines up with the flat face of the port, and that the bolt-holes are correctly aligned at the top.

Glands, Parts J and K (Fig. 5)

These are straightforward chucking jobs; you can machine the little plug to a snug fit in the holes in the covers. Mark out in the chuck as described for the cylinder-covers, but before drilling, use dividers to mark out the radius of the oval. Drill, and spot through to the cover and steam chest respectively, and drill and tap these. File up the outline, and the job is done. These are rather small and fragile parts, by the way, so go easy on the vice and chuck whilst gripping. Incidentally, the reason for the double diameter on the body is to give the impression of a normally stuffed gland despite the fact that it is an 'O' ring and needs no adjustment. Note that 'O' rings are supposed to work a little in their cavity, endways, though there is about .002 in. 'nip' on the diameter.

Piston, Part F (Fig. 5)

You need to make the piston-rod first, Part BA, but this is a straight turning job and needs no description, except to use the usual tricks of cigarette paper in the chuck to hold the stock true. Now, take the piston casting and set it flat on a chunk of steel, and *lightly* hammer it all over, using the crosspeen hammer, about 4 oz—a tack hammer. Lots of rapid, light blows, like a woodpecker, on both

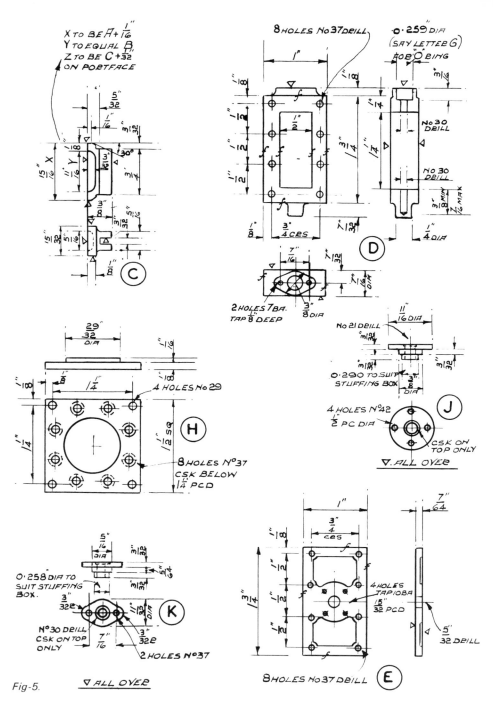

Fig-5.

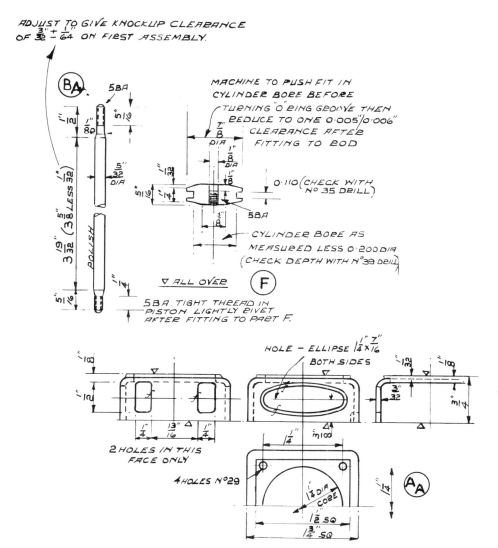

Fig 5-(Also opposite). Details of bottom cover, pedestal, piston, and valvechest cover.

sides, and on the edges. Go over it 3 or 4 times. This process 'closes the metal' or work-hardens it as we should say nowadays. Now grip in the chuck and trim the chucking-piece, at the same time facing that side. Reverse, and bring to diameter plus about 10 thou. and face to dimension. Rough out the groove (this doesn't mean being careless with it—just leave a few thou of metal in) and then centre and drill to tap 5BA, opening the hole to .125 in. about $\frac{1}{8}$ in. deep. Saw or part off the chucking piece, and face off any roughness.

Chuck the piston rod right end outwards, and set dead true, using packing as required, and screw on the piston really tight. Lightly skim the faces if they are

23

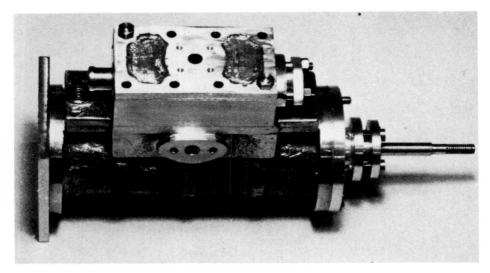

Fig 6-The cylinder erected to check piston and rod clearances.

more than .003 in. out of true, and bring the O.D. to a snug fit in the cylinder. Machine the groove to dimension—deep enough to take a No. 39 drill when holding a straightedge on the flanges, and wide enough to accept a No. 35. The finish must be good. Now skim .005 in. off the diameter to give the necessary clearance. (This is to allow for the fact that the piston will 'wag' a bit, as the parallel motion never gives a truly straight line to the crosshead.) Remove the sharp edges from the flanges, but don't 'radius' them—just ease off any burrs.

If you wish to use soft packing instead of an 'O' ring then it is only necessary to machine the groove to $\frac{1}{8}$ in. wide and the same depth. Soft packing is quite satisfactory—once run in the engine will need no more than two or three lbs/sq.in. steam pressure—probably a lot less!

You will have noticed that part BA is subject to adjustment to allow for any cumulative errors in other parts; for example, if the columns are made a shade short, then the piston will come *twice* that error nearer to the bottom cover. The correction for these is made by adjusting the length of the rod, but this can be done with the piston in place, as the majority of chucks will accept the piston behind the chuck jaws. Such an adjustment can also be made by fitting shims in appropriate places—under the cylinder, for example.

Pedestal, Part AA (Fig. 5)

This need take very little time. To reduce cost, the oval holes are not cored through, but all that is necessary is to mark out, drill holes and then file to the outline. This should leave a small rebate all round the hole, and provided the 1$\frac{1}{4}$ in. dimension is held it will be best to work for appearance rather than dimension. The faces may be machined in the 4-jaw, but care should be taken first not to crush the relatively thin casting and second, to ensure that the height is uniform; it may be necessary to do a bit of selective filing to get this right. The holes for the fixing bolts are marked off from the cylinder cover. Incidentally, I have used a shade larger clearance drill than usual, so that the exact position can be adjusted on assembly.

This completes the cylinder set; it has taken rather a long time, but if this part isn't right, the rest of the engine

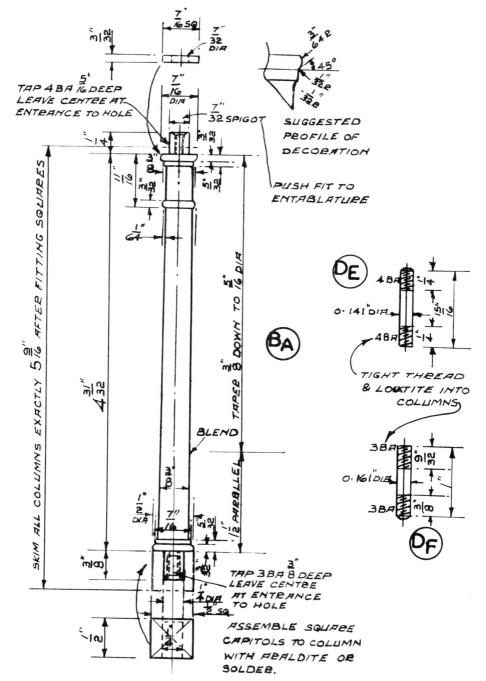

Fig 7-Details of the columns. The decorated rings can be made as separate pieces—see also Fig 9.

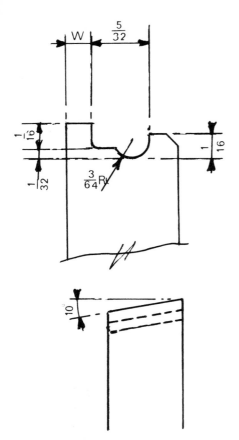

Fig 8-Form tool for machining decoration on column. Dimension 'W' not less than width of parting tool.

can be scrapped! I do not recommend the fitting of studs etc. at this stage—just assemble with a few screws to keep all together. Fig. 6 shows the completed cylinder.

Before going any further, a note about the length of studs shown in the schedules and drawings. These are given in the normal 'real' engineering way, —that is, the amount of projection when the stud is installed called the 'standout' or 'face-to-point' dimension. Some suppliers quote the *full* length overall, and this point must be watched. Further, it may be that standard studs available are not exactly the right length. If this is so, then they must be shortened. Note that studs should come hard up against the end of the thread at the surface of the casting and *not* bottom in the hole. The normal engagement of a stud in its hole is the stud diameter plus two threads. You can use Loctite if you like, but it should not be necessary.

Now, I was going to go straight to the other end of the engine and make the crankshaft, but I had the machine set up for taper turning at the time and as there are two parts on 'Mary' which need the tailstock set over I thought these might as well be done first. You can then reset the tailstock dead true—as you must for the crank—and leave it for the rest of the time.

Columns (Fig. 7)

The drawing shows the decoration turned out of the solid column, but it may be easier to make these as little loose washers; the column can then be machined from $\frac{3}{8}$ in. bar. Make a new sketch, showing the $4\frac{31}{32}$ in. dimension as $4\frac{21}{32}$ in., and the two spigots each $\frac{5}{32}$ in. longer than shown, to allow for the 'removal' of the decorated collars. Clean up the stock, and part off four pieces to exact length ($5\frac{19}{32}$ in.). Set truly in the 3-jaw, using cigarette paper if need be, turn down the spigots at the ends, centre deeply, and drill for the tapped holes. The centre must be present after drilling. Now turn the top end to $\frac{11}{32}$ in. dia. to give the top diameter of the band, and mark the position of this. Also mark with a felt pen a point about $1\frac{1}{2}$ in. from the bottom end, to indicate the extent of the parallel part of the column.

Set over the tailstock towards you by trial to give a change in *diameter* of $\frac{1}{16}$ in. between the pen mark and the position of the band and, with the stock between centres, turn the taper under power feed till the tool just brushes the felt-pen-mark. With the same setting of the cross-slide, turn the short part above the band. The next step would be to pass over to the other job that needs the tailstock set over (the eccentric rod, described in the next section, page 28) but let us assume you have done that and reset the lathe to turn parallel. With the column again between centres, blend the parallel and the taper portions, using, I suggest, a No. 4 cut flat precision file. Note always in using a file in the lathe that this is still a cutting tool, and has the same limitation as to cutting speed as a lathe tool—and files are carbon steel, not H.S.S! Run the lathe slowly. At the same time, very lightly radius the edges of the decorative band. Finally, with very fine emery paper 'frost' the surface; the columns should be painted finally (though it may break your heart to cover up such fine brasswork!) and you need a keyed surface for the primer.

The collars at the end may be turned using handtools, or you can make a little forming tool out of silver steel flat. Fig. 8. The easiest way is to drill a $\frac{3}{32}$ in. hole—start with a No. 43 and follow with $\frac{3}{32}$ in. to get a good finish. This hole must be drilled at an angle of 7 to 10 degrees to give the necessary clearance. Drill the hole first, and mark out the rest from it, file to shape taking care to get a fine finish. You can only sharpen such a tool on the top, and any filemarks on the profile will show on the workpiece. *Don't* use emery—this will radius off the cutting edge. Harden and temper to straw, lightly clean up the top on the flat of the grinder, finish with fine oilstone and finally, if you have one, with an arkasas stone. The tool must be used exactly at centre-height. It will take a few hours to make, and you might well have done the turning by hand in the time! The collars are made from brass stock, feeding in with the form tool till the radius just brushes the surface, when the other dimensions will

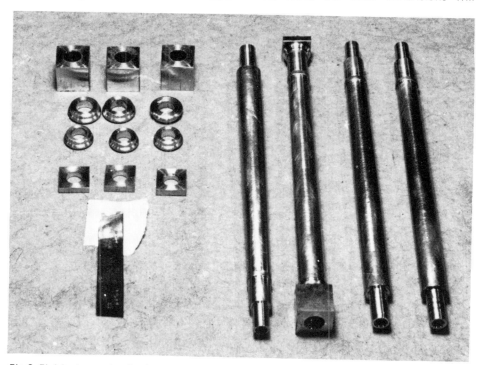

Fig 9-Finished parts for the four columns, one assembled.

be right. Face off the end slightly concave, drill to correct size, and part off; four will be made from $\frac{1}{2}$ in. material, and four from $\frac{7}{16}$ in.

For the square ends, I simply chucked square bar in my self-centring four-jaw, drilled, and parted off but if you haven't got one, use the independent chuck. Don't grip so hard as to mark the surface. You may have to file or mill the $\frac{7}{16}$ in. square stock from $\frac{1}{2}$ in. stuff for the top squares. Now tap the holes in the ends of the columns, taking more than usual care to get the threads upright in the hole. Degrease the whole issue in Carbon Tetrachloride (or soap and water) and assemble with Araldite or Loctite 601 using temporary screws and washers to hold all together and the square ends true to each other. Allow to cure, and then check the overall length. It is fairly important to hold the $5\frac{9}{16}$ in. overall length, as any error here will be doubled at the piston end, but it is even more important that all be the same length. If there is any difference, set up between centres again and skim the ends of the squares. Remove all burrs. The studs, Parts DE and DF are straightforward, but note that the depth of engagement in the columns is greater than 'standard'. Fig. 9 shows the bits. Now for the other set-over-tailstock job.

Eccentric and Rod, Parts A to AA (Fig. 10)

Rod, part D. For this, we are going to take advantage of the fact that a beam bends in a curve! Part off a length of $\frac{1}{4}$ in. steel $6\frac{1}{2}$ in. long, and centre the ends. Mark off with a felt pen at the centre, at $\frac{3}{4}$ in. either side of the centre, and at $2\frac{5}{8}$ in. either side of centre. Grip in the 3-jaw, setting truly with paper, and turn the ends—to the mark—to $\frac{11}{64}$ in. diameter. Set over the tailstock towards you about 40 thou.; grip the stock in the 3-jaw with one of the $\frac{3}{4}$ in. marks sticking out just sufficient for your tool to reach it without fouling the jaws. Set the tool *just* to touch the mark, note the index reading and traverse to the $\frac{11}{64}$ in. dia. Adjust the tailstock so that when engaged with the centre the workpiece again just touches the tool. The stock will now be bent into a nice curve! This method is hard on the tailstock centre, so use a good grease (I use a mixture of tallow and flake graphite, but the latter is not easy to come by these days; Molyslip grease or Rocol HMP grease will serve.) Work on power traverse—feeding away from the chuck if you are nervous!—taking cuts not above 10 thou. at a time and keeping a careful eye on the tailstock centre for overheating. The use of lashings of cutting oil will help cool things down. For the final cut, hone the tool and use whatever technique you are accustomed to to avoid chatter. (Mine is dead sharp tool and minimum overhang; if the latter has to be large, stick a bit of putty or plasticene on the top of the tool near the point.)

Reverse in the chuck, and treat the same way, till you just brush the $\frac{3}{4}$ in. mark on the other side of centre. Reset the tailstock to turn true. Set the rod between centres and blend the parallel part with the curve as in the case of the columns. Finally, turn down the ends, cut the threads with the tailstock die-holder, polish all if you have a mind to, and part off to length. To hold the first parted-off end in the chuck, after it has lost its centre, simply put a split 5BA nut in the jaws. As the parting tool goes through, hold the workpiece in your hand (loosely) to avoid damage.

Rod End, Part E

This is best made from a piece of $\frac{5}{16}$ in. square stock, brought to size later. I suggest you make the $\frac{3}{64}$ in. slot first, and then, if this goes awry—as well it may—you have wasted as little work as possible! To make this, mark out and drill a hole in the centre of the slot, say No. 57. File or turn a little pin to be a *tight* fit in the hole, knock it in and file off the top. Now drill two more

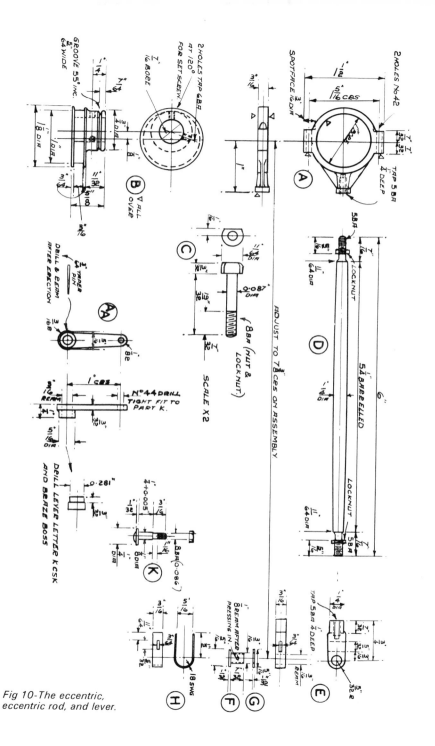

Fig 10-The eccentric, eccentric rod, and lever.

holes, one on either side overlapping the first. Remove the pin (just like that! You may have to make a tiny punch to do it!) and with a swiss file, open out the slot to size. I found it necessary to do a bit of work with a piercing saw first, but usually I find a No. 2 or No. 4 cut warding file will cope quite happily with these small slots. Mark out the dimensions from the slot, set up in the 4-jaw and turn the end, drill and tap. Mill or file to width, and mark out for the $\frac{3}{16}$ in. hole and for the radius on the end. Drill and ream, and finally form the end to shape. The little **Bush,** Parts G and F, are a straightforward turning job, the only point to note being that the washer G should be a tight fit on the bush. Assemble into the rod end (part E) with Loctite and, when cured, finally ream to size. The **Pin**, part K, is also straightforward as is the lever in which it fits, part AA. The only points to note are (a) that the pin should be a good fit in the No. 44 hole—it may pay to drill No. 45 first, in case your drills cut large; and (b) there may be a little difficulty in getting a clean letter K hole in material as thin as $\frac{3}{32}$ in. There are two ways of getting over this—first, to drill into a block of identical material below the lever; or to drill undersize and ream.

Strap, Part H and cotters

Making the strap is very much easier if you have material soft enough to prevent 'rebound' after bending. The thickness is not critical, and if you have a piece of the old-fashioned blue-planished 'steel' (often swedish iron) this is fine. Otherwise I suggest you anneal a piece of mild steel—simply heat to a dull red or a little over and allow to cool slowly—it won't hurt to quench it, preferably in oil, once it is down to black. It should now bend much easier. Cut to size, a trifle over the length (say $1\frac{9}{16}$ in.) and bend to shape. Offer up to the block E and mark the position of the slot; *note carefully* —the slot in the strap should be $\frac{1}{32}$ in. nearer the radiused end in the strap than in the block, so that the cotter will pull it all up tight. Drill a couple of holes, open out with a piercing saw—even a metal-cutting fretsaw will do in this thin material—and finish with a file. Bring to length and remove burrs. The cotters are the same as those for the connecting rod, so these may be left till later—see drawing Part AH. The bolts, part C, are simply standard 8BA with modified head, but you can make them from scratch if you like.

Eccentric Strap, Part A

This is made in the usual way—trim the casting, drill for the bolt-holes, cut in two, mark the mating parts and trim the joint, bolt together and bore in the 4-jaw and face the sides, and so on. (There are 'n' other ways, of course, but thats the way I did it.) The one point that does need attention is the drilling and tapping of the hole for the eccentric rod. This must be true and square or it will both look wrong and work badly. The way I did it was like this; after finishing all other work, set up a vice on the vertical slide and adjust it till the rod half of the eccentric strap is a push fit into the vice. Fit the headstock centre and advance the saddle till the point enters the jaws of the vice; adjust the vertical slide till the two jaws both touch the centre. The gap in the vice is now true to the lathe centre. Grip the half-strap, with the bolting face hard onto a piece of packing. Then if the slide is square to the machine centre, so is the work. Now adjust the saddle till the headstock centre is at the centre of the boss on the strap. Centre and drill No. 37, then with a little endmill face the end to bring the 1 in. dimension correct. Tap 5BA. I usually like to relieve the bore about 20° either side of the joint-face, by about .005 in., which prevents the risk of the strap from closing in on the sheave, and also provides a wedge for lubrication. No oilhole is shown—it isn't really necessary, but you can put one in if you like.

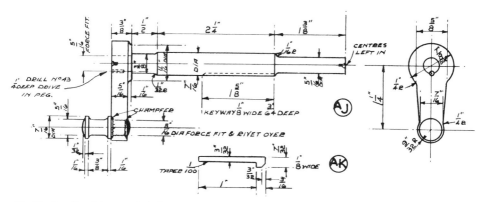

Fig 11-Details of the crankshaft.

The **eccentric**, part B, I propose to leave till the crank has been made, as although it is the reverse of good (full-sized) engineering practice, I find it better on models to bore holes to fit shafts rather than the reverse. Provided you DO bore, and don't ream!

Crankshaft, Part AJ & AK (Fig. 11)

Having said which, we will now proceed to make a couple of holes first, and fit the shaft afterwards! The web is made from a piece of M.S. or stainless flat, 1 in. × $\frac{3}{8}$ in. × $2\frac{1}{8}$ in. long. You can either set it up in the 4-jaw or use the drilling machine; the latter saves time provided it drills square. Mark out for the two holes and with dividers, draw a circle $\frac{5}{16}$ in. dia. and another slightly smaller—say $\frac{1}{4}$ in.—inside it, and the two circles $\frac{7}{8}$ in. and $\frac{9}{16}$ in. dia. for the radiused

Fig 12-'Shaping' the radius on the crankweb in the lathe.

Fig 13-The crankshaft completed.

ends. Then the two lines for the flanks of the web. Enlarge the centre-pop with a spade-bit in your watchmakers drill (posh name for the old fret-saw archimedian drill!) and then very cautiously open it further with say a No. 35 but don't go any deeper than the point. Grip in the drilling vice, and start drilling letter N. When the point nearly reaches the $\frac{1}{4}$ in. circle, have a look and see if it is central. If not, make a little groove down the conical hole, right to the point, on the side towards which you want the drill to move. This will 'wander' the drill; have a second look when the drill has cut out this groove, and repeat if need be. Then take the drill right through. Repeat for the other hole. Chuck a $\frac{5}{16}$ in. reamer, use ample cutting oil, oil the machine table so that the vice can slip, and ream the hole in the machine. For a H.S.S. reamer 450 rpm is about right for M.S., but clear the chips *very* frequently. Lightly countersink the holes. Hack off the greater part of the surplus metal from the profile with a saw, and then bring to outline with a file. To get the radii right you can plane the shape in the lathe, as shown in fig 12. The work is mounted on a mandrel—stub or otherwise—with clearance behind, and the tool mounted at exactly centre-height. Zero or even negative top rake. After roughing out this way, filing to shape is easy.

Whilst on the mandrel, the $\frac{11}{16}$ in. dia. boss can be formed, and the thickness reduced to $\frac{5}{16}$ in. The rear of the hole for the crankpin must then be lightly countersunk for the rivetting. Remove all burrs. The shaft may now be turned, between centres, from $\frac{1}{2}$ in. stock. *Don't* use ground mild steel—even the engines at the great Exhibition of 1852 didn't have ground crankshafts! Make the diameters to micrometer dimensions, plus 0 to minus .001 in. on the bearing seats ($\frac{3}{8}$ in. and $\frac{5}{16}$ in. at the long end) and plus/minus .001 in. on the $\frac{7}{16}$ in.—this doesn't mean that a taper of a couple of thou. is permissible—it isn't! But the diameter doesn't matter to within these limits as you will make the flywheel and the eccentric to suit. The $\frac{5}{16}$ in. stub that presses into the web should be made between $\frac{1}{2}$ and 1 thou. interference. It is best to turn this diameter first, after having roughed out the rest, and then if you make it too small you can start again. Aim at a good tool finish, try and avoid having to use emery if you can. Yes, free-cutting steel is quite O.K. for the shaft—the original would have been wrought iron.

The crankpin can be turned from the chuck, the same remarks applying as to the diameter of the part that presses in. Leave this about $\frac{1}{32}$ in. too long to allow for rivetting. Press the pin in first—it will be easier to get the hammer at it. If it is

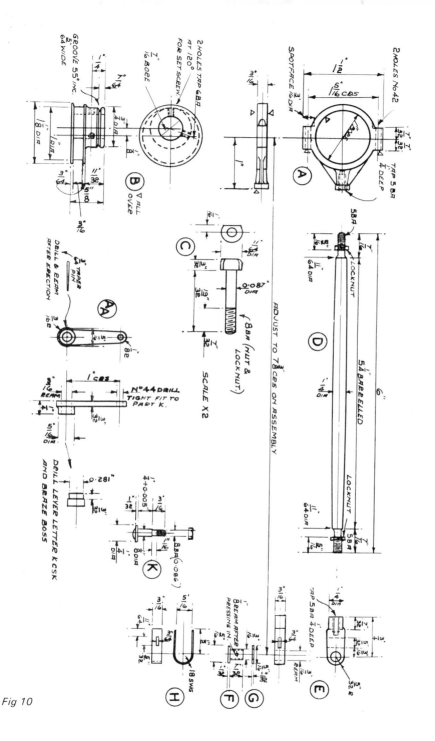

Fig 10

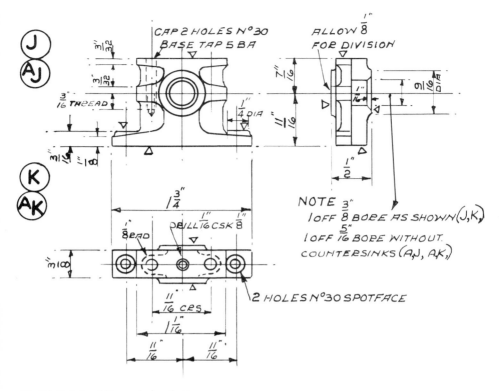

Fig 14-Details of the main bearings.

too tight, it is quite in order to rechuck and ease with emery or a No. 2 cut swiss file. Or you can heat the web. Rivet with care, as this part will show, and you can do little about stray hammer marks on the web. Now press in the shaft; It will probably be too big for your vice—it was for mine—but I have one of those remarkable hammers supplied by the THOR people which are filled with lead shot and have a nylon or plastic head. This sorted it, without any marks on the end either, by interposing a bit of soft lead between hammer and work. If the stub comes through too far, you can remount between centres and turn it off flush. Drill for the little peg, getting it in line with the crankweb centre, and then set up on vee-blocks in the vertical slide and mill the keyway with a $\frac{1}{8}$in. endmill—getting this, too, in line with the crankweb. The key can be left until the flywheel is made—it is a straightforward filing job and needs no instruction. Fig. 13 on page 32 shows the completed crank.

Eccentric Sheave, Part B (Fig. 10)

This part can now be made, either in the traditional way by setting over in the 4-jaw, or by fitting packing under one jaw of the self-centring chuck. The hole should be bored until it is a sweet sliding or mild push fit on the shaft. At the same setting, machine the back—the flat face—of the sheave, or you can mount on a mandrel if you wish. This gives a reference face to bring hard up against the back of the chuck (with packing if needed) to ensure that the working surface is true to the bore.

Machine this until the eccentric strap only *just* fails to meet when offered up to the groove, thus giving a margin for bedding in with scraper later. Put back on the mandrel and turn the boss on the front face, and make the groove for the drive belt to the governor. The two holes for the grubscrews can now be drilled and tapped—it doesn't matter if they are not spot on at 120°. Note that when the eccentric is fitted to the shaft, you should very lightly pinch one of these screws to leave a mark, and then set up the crank between centres and turn a groove about $\frac{1}{16}$ in. wide and $\frac{1}{64}$ in. deep to receive the point of the grubscrew. This will avoid raising a burr on the shaft. These set-screws on the original had little square heads, which you can make by filing down a standard 6BA hexagon screw.

Main Bearings, Parts J, AJ, K and AK (Fig. 14)

Note that these are two different diameters in the bores. First, after dressing off the castings, file or mill the top and bottom faces to bring the $\frac{3}{32}$ in. and $\frac{1}{8}$ in. dimensions on the flanges correct, and to provide reference surfaces for further work. Now mark out (fig 15) and drill and tap for the cap-studs—clearance in the top cap, of course; don't go too deep with this drill. Cut in two, marking the mating halves, and bring to dimension with reference to the already machined surfaces of top and base. To ensure that both bearings are to the same centre-height you can mount the two bottom halves together on the lathe faceplate and machine at the same time. Assemble

Fig 15-Marking out the main bearings.

together and mount on the cross-slide on packing sufficient to bring the joint to centre-height (1$\frac{3}{8}$ in. on a Myford, but check this!) drill and ream. Mount on a stub mandrel and finish the whole to width, but leave the set with the $\frac{3}{8}$ in. bore (Parts J, K) about 10 thou. wide, so that you can adjust the crank end-float later.

Relieve the bearing at the joint face as in the case of the eccentric strap, but *don't take the relief right up to the end of the journal*, so that oil is retained. Drill for the oil-hole (there is a little dummy lubricator—part DH on the other drawing—to keep the dirt out) and with a tiny chisel, or dental burr if you have a flexible drive shaft unit, cut an oil groove from the oil hole to the two reliefs just mentioned. Offer the bearing to the shaft. You may have to take some off the faces of the inner main bearing, parts J & K, to get it in, but don't take so much off as to make it sloppy. If the bearings are slack on the shaft, file a little off the joint faces; if tight, get to work with a scraper. Finally, dress up the exterior, drill and spotface for the holding-down studs (it won't hurt to make these holes one size drill larger than shown) and, if the original marks have been machined off, remake the mating parts right way round.

Bedplate, Part A (Fig. 16)

There are several ways in which the part can be handled, but the first step must be to trim up the casting, removing all flashes and roughness. Now file the underside to get a flat surface to sit the thing on during later processes, bringing the thickness of the lip round the base the $\frac{3}{32}$ in. I use the table of my drilling machine as a surface plate for jobs like this, but a good alternative is an old mirror; you can get a complete wardrobe or dressing table at the local auction sale for as little as 50 pence,—less than you would have to pay for the hinges—and this will give you both a good surface plate *and* a supply of good, well seasoned wood as well.

Now for the top. You can mill it in sections, flycut it, or file it. (If you have a hand shaper, this will serve better still.) I prefer the last method, but will deal with the others first. For either you will need to hold the base on an angle-plate or vertical slide, and this means some holes. Drill the 6 holes in the holding down bosses, centring them to the casting—the exact holding down position doesn't matter all that much. You can also drill holes—say $\frac{1}{4}$ in. clear—in three places without them showing in the final job, though, of course, you can always plug afterwards wherever they are. The holes may be positioned on the cylinder centre, on the pump centre (a $\frac{17}{64}$ in. hole is needed here anyway) and on the beam centre-line—this one will be underneath the governor bracket in the end. These holes, together with that provided in the casting to clear the crank, will give sufficient clamping provided you confine yourself to fine cuts and use a sharp cutter, but if you have any doubts, drill some more to a convenient tapping size and plug them after you have finished. Put the casting on a flat surface, traverse over it with a scribing block to find the 'low point', and mark this with a felt pen. Set up the vertical slide, taking more than usual care to get it square; check that it is square vertically also, putting paper shims under to correct it if need be. Clamp the casting to the slide with the 'low point' in the middle of the working surface.

Note that though the schedule shows the base in Gunmetal it is, in fact, light alloy. This change reduces the cost and should make it easier to machine.

Now, using the leadscrew to apply the cut, mill as much of the surface as you can, but don't overlap more than say $\frac{3}{4}$ in. beyond the edges of the slide. Mark the leadscrew handwheel when you make the final cut. You will have to work round the nut on any clamp bolts, but the protuberance will be easy to file off afterwards. Reset the casting in a new position, and repeat the operation, using the mark on the handwheel to set

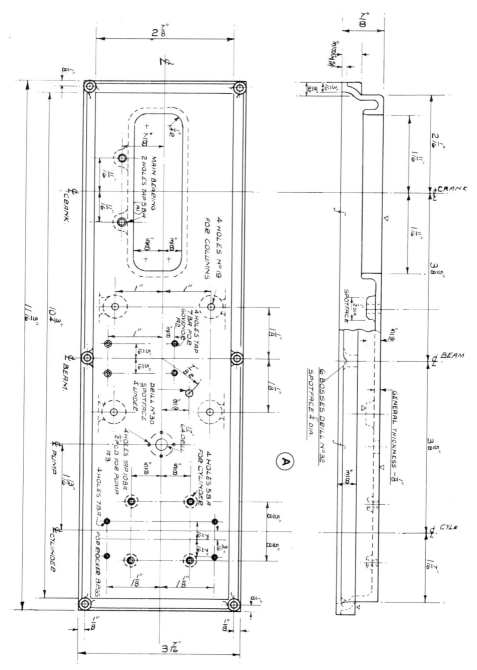

Fig 16 - Details of the bedplate.

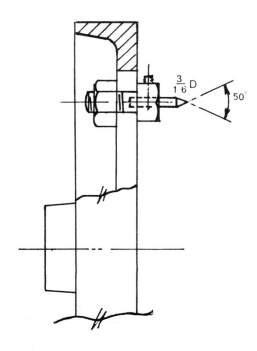

Fig 17-Improvised flycutter mounted on the lathe faceplate.

for the last cut. And so on, till the whole is covered. You may well find that at one setting you can machine off the boss left under the nut used in the previous setting over some parts. You may think that this is a very rough and ready way of doing things, but in fact it is common practice in jobbing machine shops, and often the only way that big repair jobs can be handled when the planer or milling machine isn't up to the job.

Flycutting involves a similar set-up, but here the face can *just* be covered in one setting. Holes will be needed as before, but should be countersunk, and you will need one on either side of the crank cavity also. The first step is to find a chunk of dry wood a little larger than the casting base—trying the method out I used a piece off the end of a 4 in. × 4 in. oak fence-post, about a foot long. The cutter is made from a piece of silver steel set in the end of a bolt attached to the faceplate—fig 17—though you can use a piece of round H.S.S. if you have any. Set this at the largest radius available on the faceplate—make sure it just clears the bed in the gap. Drill through the wood to accept holding down bolts—or use clamps, though through drilling is better—and attach this to the cross-slide. Set the block as square as you can, and then machine the face of the wood with the flycutter. You will get a bit of interference, when the cutter is cutting both at the back and the front part way across, as lathe cross-slides are always set to machine very slightly concave. This doesn't matter. Smooth off the surface with an old file or with sandpaper, and attach the casting to the block using countersunk *brass* screws in the face, and steel ones in the 6 bosses. Make sure the screwheads are well below the level to which the top is to be machined, or you will chew off the slot in the heads! Run at about 50 rpm or so (or 100 rpm for H.S.S.) and use a very slow feed, again putting cut on with the leadscrew; in this case however, it is most important to lock the saddle at each cut. Further, after each cut till the final one, check that the screws haven't been jarred a bit loose by the shock of the interrupted cut. Again, you will find a trifle of interference, but not to worry, as you will equally find the finish not too good, and will have to trim to a finer one with a file. You can improve the finish by fitting two such cutters, but must take care that the cutting faces project by equal amounts—check with a dial indicator. (If you were to fit 6 or 8 cutters you would have a production style inserted tooth face mill, of course!)

As I said, I 'machined' the prototype with a file—fig 18. There is no difficulty in this provided you have the correct tool for the job, which is a 12 in. or 14 in. second cut 'hand' file. Dose it well with chalk to avoid 'pinning'. A professional would start with a bastard cut and finish with a smooth, but amateurs are likely to get chatter with anything rougher than a second cut. The file must not have been used on

Fig 18-The bed can be 'machined' with a file as shown here.

steel. The secret of flat filing is first to get a comfortable stance, with the work at a convenient height; to use the full length of the file with a slow stroke—files are carbon steel, and at 100 strokes/minute with a 12 in. stroke the cutting speed is nearly double that appropriate to *high speed* steel!—and to check frequently for flatness. The file is *not* flat, but slightly convex, and this will, if you let it, correct for the inevitable curvelinear movement of your hands. File diagonally across the surface, working from one end to the other, then on the opposite diagonal back again, occasionally giving a few strokes down the centre of the work from one end to the other. Towards the end you may find it helpful to use marking blue and take down the high spots with an 8 in. file, but I think you will find that the light reflected from the filed surface will tell you if you are getting the work at all 'bulgy'. Keep this file well charged with chalk and clean frequently to avoid 'pinning'.

Whichever method is used, it can be noted that the $\frac{7}{8}$ in. height of the bed is not all that important, provided you make the outer bearing support, part B, to match. The surface should not be polished, by the way, as the bed is to be painted, and a draw-filed finish will give a good key to the paint.

The bed may now be marked out and the various holes drilled and tapped. I suggest you leave those for the Pump and the Governor-bracket, to be marked out from these components in due course, but don't forget to put the centre-line for the latter. The bosses below the column holes on the underside require to be spotfaced to the $\frac{5}{16}$ in. dimension shown. The final operation is to smooth off the roughness from the decorative moulding, but again, don't aim at a polished finish.

Outer Bearing Support, Part B (Fig. 19)

As in the case of the bed, trim off the casting and file the underside till it is flat, and the lip comes to $\frac{3}{32}$ in. Chuck in the 4-jaw, taking care that the casting is seated truly on the face of the chuck, and machine to height—this to correspond give or take 5 thou. with the finished height of the bed. The casting is pretty robust, but take care for all that not to crush in the sides. Mark out for the holes for the bearing and drill and tap. Those for the holding-down bolts should be centred on the bosses and spotfaced.

Entablature, Part F (Fig. 19)

Great care must be taken in chucking this casting, as it will be found very easy to distort. Set in the 4-jaw, top outwards, and take a trial cut across the face. Now check that the flange, shown as $\frac{5}{64}$ in. thick on the drawing, is uniform; if not,

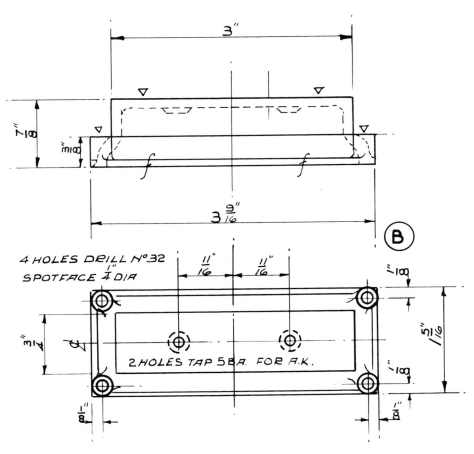

Fig 19-(Also opposite page). Details of entablature and outer bearing pedestal.

adjust in the chuck, and then machine till the flange is to dimension. Reverse in the chuck, gripping the inside, and make sure the previously machined face is down flat on the jaws. Machine to $\frac{11}{16}$ in. THIS DIMENSION IS IMPORTANT, as any error here (as in the height of the columns or bearings) will alter the clearance at the ends of the stroke. Remove from the chuck, and check the height at each corner with a micrometer. If the dimensions are not within a thou. or so of being the same, correct with a fine file.

Mark out for the holes (fig 20) but before drilling those at the corners, draw a circle on each center $\frac{3}{4}$ in. dia., and a second one inside this $\frac{9}{16}$ in. dia. Lines scribed touching these circles will enable you to trim off the outline of the top to dimension. Drill the holes No. 26, but take care. Hold the work in the machine vice, and keep control of the feed. The drill will tend to take charge in gunmetal unless you are using either a straight flute drill, or one with the cutting edge stoned to zero rake. Drill and tap the holes for the bearings, and then trim the casting to outline. Take care over this, as the part is very prominent on the finished engine.

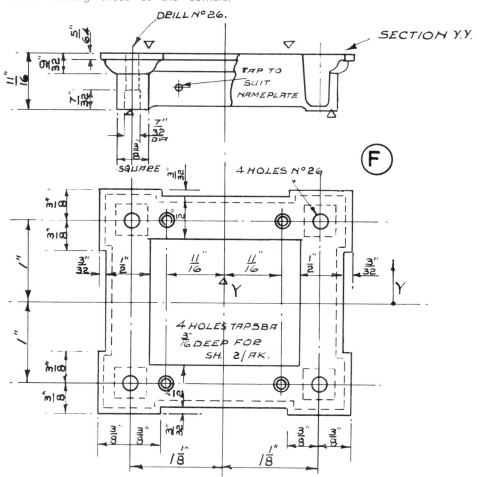

Fig 20-Marking out the entablature. Note the small archimedian drill used by the Author when starting a centre-pop.

Rocker Bearing and Cap, Parts C and D (Fig. 21)

This is a simple part, and scarcely needs description. I suggest you start by trimming the casting to the correct thickness first, though, as this simplifies later work. Saw in two, and face the sawn surfaces with a file, bringing that for the base to the correct $\frac{3}{8}$ in. and $\frac{3}{32}$ in. dimensions by filing both base *and* sawn surfaces. Drill the hole No. 24 and then ream $\frac{5}{32}$ in., or if you haven't a $\frac{5}{32}$ in. reamer, just poke a $\frac{5}{32}$ in. drill through after the No. 24. Drill for the cap bolts, with the cap still soldered on, tapping size first, and then as far as the join with No. 37. Tap with the cap still on, as this helps to hold the tap straight. Number the mating parts, and unsolder. Finally, drill for the two fixing bolts and spotface. No dimension is given for the oil-hole, but this can be $\frac{1}{16}$ in. slightly countersunk. Note that on one bearing the $\frac{1}{4}$ in. width acts as a location for the rocker-shaft. This can be filed—or spotfaced—to suit after the shaft is machined. In passing, do *not* remove all the solder from the joint faces; leave this till later, as you may have to adjust the bearing cap to fit the shaft.

Interlude

You have got a fair bit of work done now, and I suggest you give yourself a boost to morale by offering up the parts you have made. This will also have the advantage of showing up any slight errors which may have crept in. Check, for example, that the entablature drops onto the columns reasonably sweetly—if it doesn't, the holes may not be correctly centred, or the bed may not be dead flat where the columns bed on it. Check also that the entablature top is parallel to the bed, and if not, correct the lengths of the columns. And so on. But chiefly, stand back and look at the job; I am pretty certain that you will regret that you are going to have to paint it!

Beam, Part AJ (Fig. 22)

The casting is provided with four little projections on the flange of the beam to enable it to be clamped without strain. Dress off the casting and file the pads on one side to line up so that it sits on a flat surface without rock. Make sure that the casting is reasonably true to the flat surface—the beam will be on the skew otherwise. Check that the casting itself isn't twisted, as these slender gunmetal ones do suffer occasionally. It is safe to 'bend it straight' if the twist is slight. Set up on the faceplate, true to the centre boss, and clamp adjacent to or on the support pads. Machine the faces of all the bosses to the same final setting, so that they are level to each other, Fig. 23. True up the exterior of the centre boss, and then drill and ream to $\frac{3}{8}$ in. bore. Take from the faceplate, and file the pads on this side to lie level with the machined boss faces. Set up on the faceplate again, true to the bore in the centre boss, and machine the faces and the outside of the centre boss. Whilst still in the lathe, mark out the centre-line through the bosses and, with the beam vertical, for the centre distances of the holes. It is more important that the holes in the two end bosses be the *same* distance from the centre than that they should be exactly $3\frac{3}{4}$ in.; so if an adjustment of up to .010 in. will bring the hole nearer the boss centre, no harm in making the adjustment. On the other hand, it *is* important that the $1\frac{7}{8}$ in. dimension between the hole at the end of the beam and that for the parallel motion be held true to drawing, so don't adjust this one.

In marking out these accurate distances, I scribe as carefully as I can, then put a *very light* dot at the intersection of the lines. This I enlarge with a watch-

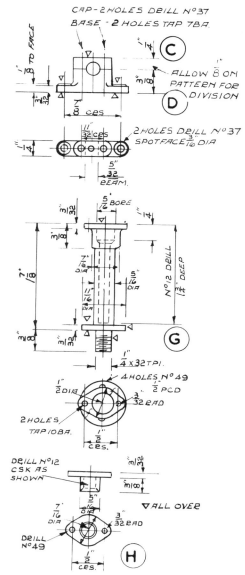

Fig 21 - Details of rocker bearing and pump body.

makers spear-point drill—like those for fretwork, but very hard and sharp—used in an archimedian drill. By examining with an eyeglass it is possible to get a really accurate and quite deep centre, which can then be further enlarged with

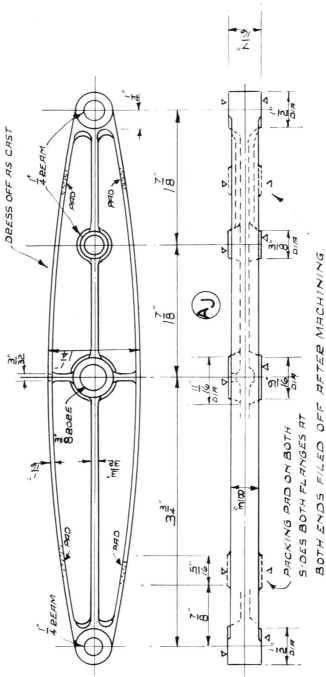

Fig 22-Details of the beam.

Fig 23- Machining the beam. The tool must be retracted to clear the clamps when facing the bosses.

a slocumbe bit. These watchmakers 'forets', as they are called, can be obtained from horological suppliers, in various sizes. (Thomases, who supply the pins for this engine, list them) I use the No. 8 size, which is 0.8 mm dia.

Take the beam from the lathe, and set flat on the drilling machine table, having first removed any machining burrs. If there is any rock, seek the cause and remove it. Taking the same care as mentioned earlier to avoid the drill taking charge, drill the bosses letter D and ream $\frac{1}{4}$ in. Run at about 1000 rpm for the reaming operation, and work like a woodpecker—frequent withdrawal of the reamer to clear chips. Lightly countersink the holes both sides. File off the support pads, and trim the whole beam to outline, then clean up the interior of the flanges to get them visually uniform.

Beam Trunnions, Part AK (Fig. 24)

These need a little care, as there are a number of centres to be kept true. After trimming off the casting, paint over with whiting or specto-blue and by trial marking-out establish a vertical and horizontal centreline that effects the best compromise between the exterior appearance of the casting, the centre of the beam pivot, the centre of the boss on the 'ear' which carries the parallel motion, and the base. It is often difficult to hold a casting whilst doing this sort of job, so try either standing it on a bit of plasticene, which enables slight adjustments to be made easily, or stick it to an angle-plate with Bostik 'Blue-tak', which is a substance like plasticene but with more adhesive properties. Having done the best you can, file the underside of the

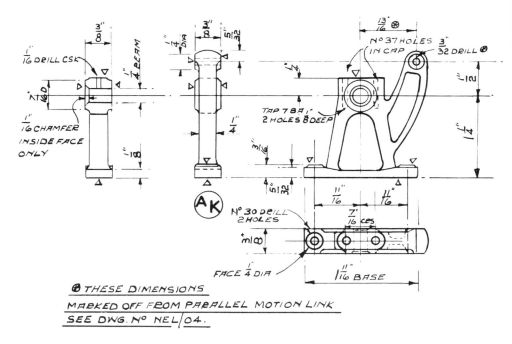

Fig 24-Details of the beam trunnions.

base flat and to the marks, to act as a reference face. Final marking-out is then done from this face. Mark out the 1¼ in. height boldly, and saw off the caps. Bring the lower half to this dimension, face the bottom of the cap, and solder together. Now file one side of the casting and the bosses flat, so that it can lie on the drilling machine, making very sure that it is square to the base. Centre and drill the hole—I suggest in stages, finishing letter D and reaming ¼ in. as for the beam. You could, of course, set the job in a vice on the vertical slide of the lathe, and do the job with drills in the 3-jaw, but though I was tempted it seemed a lot of trouble, and in the event the holes done as described proved to be quite true.

Mount the casting on a mandrel, taking care not to force on too hard and break the solder joint. You can now trim the sides of the castings—the width is the 'same all the way up'—and form the $\frac{7}{16}$ in. dia. outline of the boss. The hole should then be countersunk. Look at the drawing carefully, it is countersunk on *one* side only, but on the *opposite* side on each casting. You can now drill for and tap the holes for the cap bolts, using the same method as for the rocker bearings, and finally for the holding-down bolts. Note that though it says 'spotface' you will have to file the boss that lies under the 'ear'. The casting can then be trimmed to outline all over before marking the mating parts and unsoldering. Finally, set up on the surface plate and mark the horizontal centre-line *only* of the hole in the 'ear'—1¾ in. above the base. This hole will be marked out in the other plane when erecting the parallel motion.

Flywheel, Part E (Fig. 25)

Having made the beam it would be natural to go on to turn up the various pegs that fit it, but that means getting the machine smarmed up with cutting oil, all to be cleaned off for the next gun-

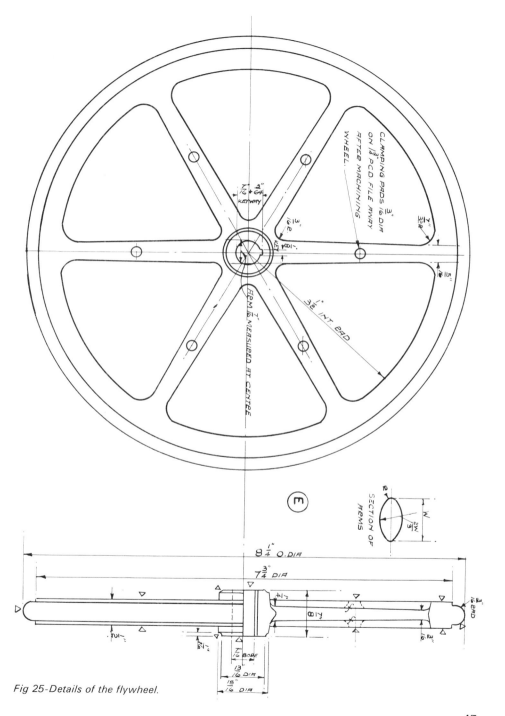

Fig 25-Details of the flywheel.

Fig 26-Machining the flywheel. **Note**—*the photo shows a 'pre-production' casting. Those now supplied have the normal machining allowance.*

metal job, so I am suggesting tackling the wheel next; but its up to you! The casting comes with little pegs on the arms, to give a bed when mounting on the faceplate. File these up on one side so that all 6 of them touch a flat surface at the same time that the rim also touches. Clamp all to the lathe faceplate, the clamps between the pegs and the rim, and arranged so that the tool will clear—you will be machining the pegs as well as the rim, by the way. *Note carefully.* The boss of the casting may project so far as to foul the end of the headstock mandrel when mounted on the faceplate. If it is likely to do so, then you must mount it on packing. But check carefully, as if you tighten clamps with the boss in contact with the mandrel you will bend the spokes. Set the wheel to run with the *inner* radius of the rim true, as this, not being machined, will show as a wobble otherwise.

Rough machine all over, including the pegs, using backgear (say 70 ft/minute) on the rim. Resharpen the tool, and fine machine, forming as much of the rim as you can reach, but leave a machined parallel part to set a dial indicator to when you turn the casting over. Finish turn the boss, but don't drill it, and finally face the pegs to align them with the face of the rim. Fig. 26 shows the casting set up, but note that yours won't have such a heavy rim. This one was the original casting from which the final brass pattern was made, and had to be packed out a bit, but you will be able to see how the clamps were fitted.

Turn the casting over on the faceplate, mounting as before, and set the outside diameter of the rim to run true, as close as you can; use a dial indicator if you have one. On the full size engine the rim would certainly have a wobble on it, as they were seldom machined, but in a model like this people will criticise even a few thousandths out of true, I'm afraid! Machine all as before, this time taking the pegs right off, as they have served their purpose. Centre the boss deeply, drill $\frac{3}{8}$ in. (keep control of the feed, as a

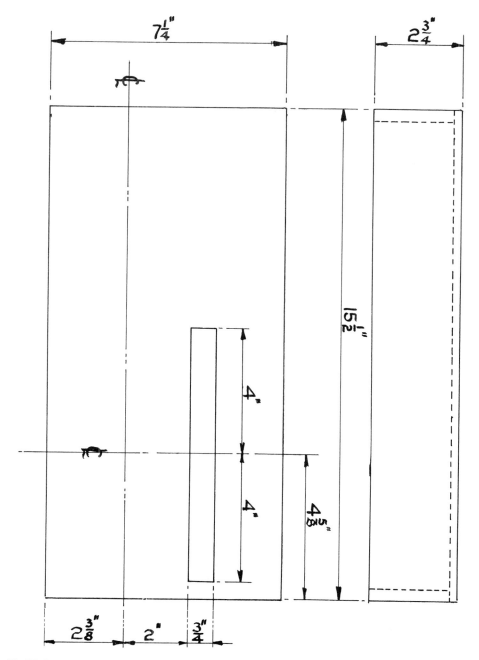

Fig 27-Details of a basebox suitable for engine erection. Top—$\frac{1}{4}$ in. plywood with ribs if required. Sides—$\frac{3}{8}$ in. deal, pinned and glued.

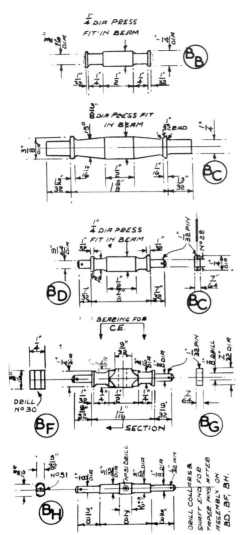

Fig 28-Details of pivot pins, and crossheads for valve and piston rod.

hole being reasonably parallel.

You may have a little trouble with chatter on the rim, intermittently, between the spokes. If so, get a hefty chunk of plasticene and plaster it on the inside of the rim to damp out the vibrations. But the real cure is to have a really sharp tool, and take a reasonable cut, don't nibble off the odd thou. The O.D. doesn't matter all that much, so long as it cleans up, and nor does the width. I finished my own rim by turning with hand tools, by the way, working by eye rather than dimension to get it looking right. Don't polish any part, as the whole ought to be painted; no full-size engine of this size would have a gunmetal wheel!

To cut the keyway, grind up a toolbit for your boring tool to .125 in. wide by micrometer, about 10° clearance and no top rake, the cutting edge dead flat. Set the bar in the tool holder with the bit sticking out sideways, so that it is central to centre-height. With this you can rack the cutter back and forth as a slotting tool in the wheel bore, gauging the depth from the feed-screw index on the cross-slide. Don't try to take more than 0.001 in. cut at a time, and rack through two or three times at each setting. The keyway depth is nominally $\frac{3}{64}$ in., but its not critical, as the key has to be made to fit anyway.

You can now take the wheel off the machine and cut off the packing pegs at the back with a little chisel. Dress off all over with a file, paying more attention to appearance and symmetry than to the given dimensions. Finish with rough emery cloth (cobblestone emery, we used to call it!) to remove all file marks, but don't polish; leave a key for the paint.

Erecting Base (Fig. 27)

We are now at a state where it will help to have something to stand the engine on as parts are offered up for trial. This is shown in the sketch. It will also serve as a guide for the final base when you get round to it. It is made up of such wood as was to hand, the top being $\frac{1}{4}$ in. plywood

normal twist-drill will want to walk through) and then bore to nearly $\frac{7}{16}$ in., finishing with a reamer. If you have a suitable gauge, better to bore to exact size; or you could bore to suit the crankshaft if you like. But most Model Engineers seem to prefer a reamed finish, and its true that you can then rely on the

stiffened with a rib, and all tacked and glued together. Dry it out thoroughly and give it a couple of coats of paint or varnish to keep it so, and thus avoid any distortion with the weather. Note carefully, the holes for the outer main bearing support should be drilled from this component when the crank and wheel are set up on the main bed, as one may have to adjust the position of these for alignment.

Beam Pivots, Parts BB, BC, BD, BE etc. (Fig. 28)

I know that there's more gunmetal still to come but if we get these next few steel parts made now you will be able to mount the beam and wheel and get some better idea of what the engine looks like.

The trick with these steel parts is to part off the material to exact length (allowing say $\frac{1}{64}$ in. for the little radius on the ends of some of them) after which you can work in the chuck, taking the dimensions from the end. However, if your chuck runs more than a couple of thou. out, turn them between centres, leaving a bit on the ends so that you can turn off the centre-hole in the chuck afterwards. They are all shown to be a press-fit in the beam, but you can use bright-drawn steel provided it is reasonably near to size. 0.001 in. over will press in well enough, and if its even twice as much as that undersize the use of 'Loctite' will deal with the matter. Purists will, of course, turn the diameter down from the next size larger stock.

The actual turning calls for little comment, as it is quite straightforward, but you should note that the two $\frac{3}{16}$ in. dia. sections on part BD don't fit anything—they are cosmetic only. You need quite a number of little collars (see the parts list) and these can be made all at once simply by drilling a piece of stock and parting off. Don't forget to face the front of each collar to get a nice finish before parting.

Part BF, the piston-rod crosshead, is made from $\frac{5}{16}$ in. square material, and is dead easy if you have a self-centring *four*-jaw chuck; I picked one up second-hand many years ago, and very useful it has been too. They are still available new, of course, but not publicised enough in my view. If not, then you must set up true in the 4-jaw, taking care not to mark the work. Centre the ends, and then work between centres. Drill the No. 30 hole before reducing the width to $\frac{1}{4}$ in., and take care that it is truly vertical and central. The cross-shaft BH is easy enough, if care is taken not to bend it, but make sure that the shoulder between the $\frac{1}{8}$ in. and $\frac{5}{32}$ in. diameters is a sharp corner.

It will pay to make all the surfaces that fit into bearings over rather than undersize, as its easy enough to open out holes, but not so easy to make slack pins a good fit! Fig. 29 shows where the bits fit.

Fitting the collars

The drawing asks you to fit and drill the collars for the taper pins 'after assembly' but this can be a woefully fiddling sort of a job—you need four hands to hold the parts still and handle the drill; a No. 65 drill at that! So, try it this way. Turn up a couple each of little collars out of $\frac{1}{4}$ in. stuff, $\frac{1}{8}$ in. hole through, and $\frac{3}{32}$ in. and $\frac{5}{32}$ in. wide, together with another having a $\frac{9}{64}$ in. hole through. Use these as distance pieces instead of the links that should normally be on the pins; slip on the collars BG, BE, etc. outside them and clamp up in the drilling machine vice, moderately tight so that they can't turn round. (Grip the shaft endways, the jaws on the collars) Carefully centre-punch, and even more carefully drill. With a No. 65 you must run as fast as you possibly can—2,500 rpm is what I used, but I would have gone to 6000 if the machine would have served me so—and withdraw the drill frequently to clear the chips. 'Frequently' means about every $\frac{1}{64}$ in. of cut. Use a tiny drop of cutting oil, and above all *feed*

51

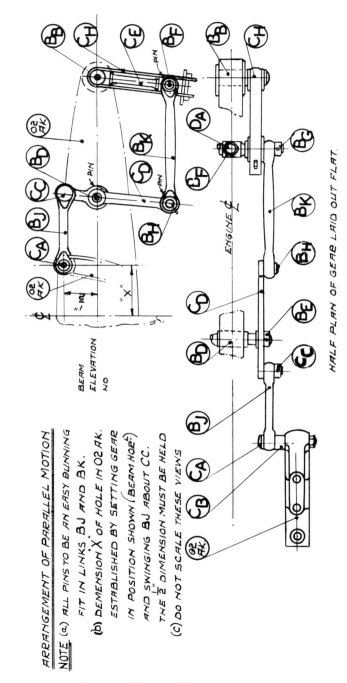

Fig 29-Elevation and expanded plan of the parallel motion.

gently, taking the utmost care on breaking through. If your drilling machine is a big one—Progress or similar—it will pay to slack off the return spring a bit, so that you can get a more sensitive feel to the job.

After drilling, broach out with the appropriate broach. Hold it in a little pin-vice, use a drop of oil, and don't force it. Broach out till the $\frac{1}{32}$ in. pin goes in half way. For appearance sake arrange matters so that the pins stand vertical on assembly, thick end upwards. Leave them assembled, as each collar will only fit in one place. I marked mine with tiny number stamps, which also show which end is the thick end of the pin. You could do the same with a tiny centre-punch.

If you haven't used any of these small drills before, then it will pay to practise a bit before you start. You must expect to break the odd drill now and then; we had a chap in the works who spent all day drilling holes .010 in. dia. in alloy steel fuel injector nozzles, and even he broke one occasionally, so you mustn't be surprised if you don't do as well! (What? Oh, yes; I broke two on the job!)

Parallel Motion, Parts BJ to CJ (Fig. 30)

Part of what follows will be equally appropriate to the valve-gear side-rods, part AF on the same drawing sheet, so I have designed the parts and the jig so that they will fit for both jobs.

Links, BJ and BK—there are two of each and they MUST be true to the lengths given—though its more important that each pair be exactly the same; that 'same' length should be exact to drawing. Further, it is vital that the holes be in line, otherwise the gear will bind. So, we will make up a jig to look after this—fig. 31.

The drill-block enables you to drill the holes central on the link, and square to it, *provided* you have the holes in the jig true. This is how to do it. Cut off a piece of $\frac{5}{8}$ in. or $\frac{3}{4}$ in. square stock and face the ends in the 4-jaw. Mark one end and one face as reference faces, and check that these are square to each other. Set up the vertical slide and machine vice on the cross-slide, and grip the piece with the long reference face towards the 3-jaw chuck on the mandrel. Adjust the vertical slide till the centre of the jig-block is at centre-height. Adjust the cross-slide so that the No. 30 hole will be in the correct position, centre with a slocumbe drill, and then go right through with No. 30. Follow with a $\frac{1}{8}$ in. drill or reamer to size the hole. Now slacken off the vice but don't on any account adjust the vertical slide. Rotate the block through 90° and set it dead square to the base (back) of the vice. Adjust the cross-slide *only* to bring the $\frac{1}{4}$ in. hole central in the end of the block. Centre, drill letter C or $\frac{15}{64}$ in., and follow with a $\frac{1}{4}$ in. drill to size. (No point in reaming it, as the stock won't be dead size to fit into it.) You now have the two holes dead in line and square. You can harden it if you like, but for the few holes we have to make its not really necessary.

The other part on fig 30 is the spacing jig, a device to get the distance between the holes in the link right. I didn't use one, as I have a 'universal' jig slab—a piece of $\frac{1}{2}$ in. thick stuff about 3 in. × 5 in. long, plastered with reamed holes all at different distances apart. To make the little jig shown, cut off a piece of flat steel say $\frac{3}{4}$ in. wide and $\frac{3}{16}$ in. or $\frac{1}{4}$ in. thick and about $4\frac{1}{4}$ in. long; these dimensions are not critical, but try and find a piece which is flat and straight. Set this in the vice on the vertical slide with packing behind to keep it square across the machine, facing the chuck with one flat side. Adjust as before to get the holes central, with the vertical slide, and with the cross-slide to get one hole about $\frac{3}{8}$ in. from one end. Set the cross-slide index to zero, or make a note of the reading.

Start the hole with a slocumbe drill, follow with No. 30, and then $\frac{1}{8}$ in. Now, use the cross-slide index to traverse over $1\frac{3}{32}$ in. and repeat; again adjust for the next hole, $1\frac{7}{8}$ in. from the first, and finally

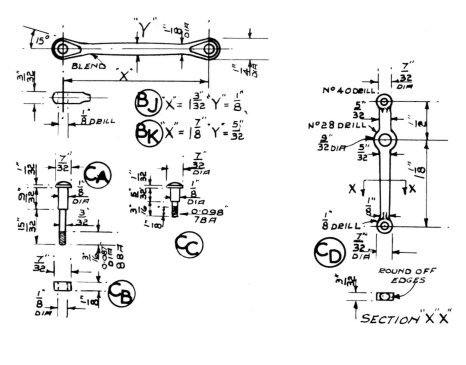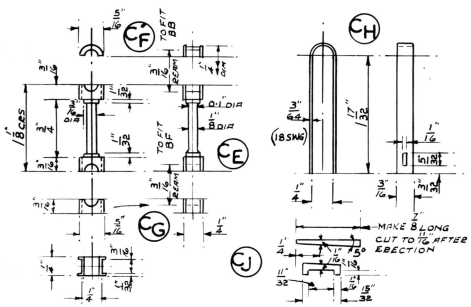

Fig 30-Parallel motion parts and other details.

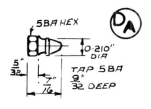
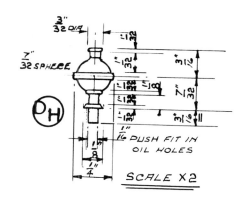
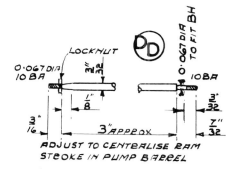
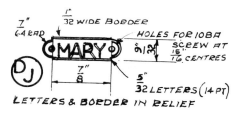
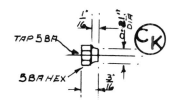
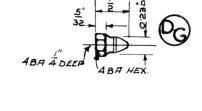
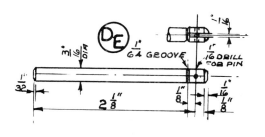
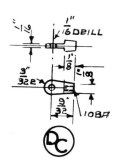

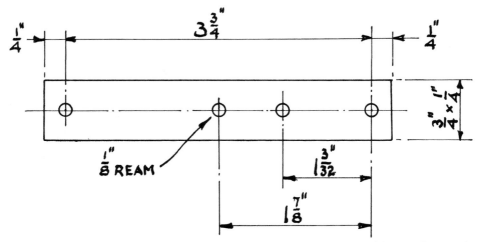

Fig 31 - Drilling jig for parallel motion and valve operating links. Two pieces $\frac{1}{8}$ in. dia. silver steel, $\frac{3}{8}$ in. and $\frac{5}{8}$ in. long also required.

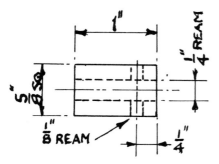

for that $3\frac{3}{4}$ in. from the first. These dimensions will be within a thou. or so of correct. The other thing you need are two little pegs—simply bits of $\frac{1}{8}$ in. silver steel, cut to length and the ends very slightly chamfered.

To make the links, part off pieces of $\frac{1}{4}$ in. stock about $\frac{1}{4}$ in. over length. (Do those for the siderods AF at the same time if you like.) Grip the drill block in the drilling vice, tapped down to be true on the base, and insert one end of one of the rods so that the $\frac{1}{8}$ in. hole will be about $\frac{1}{4}$ in. from its end. Drill the hole. Repeat with all the others, drilling a hole in one end only of each. Now set one of the $\frac{1}{8}$ in. pegs in the $1\frac{3}{32}$ in. hole, slip the block over the undrilled end of the appropriate link, and poke the other $\frac{1}{8}$ in. peg in the $\frac{1}{8}$ in. hole in the block. Put the already drilled hole in the link over the peg in the spacing jig, and adjust till the peg in the block fits the end hole, fig 32. Hold down firmly, and drill through the block into the link—right through, but don't chew away the peg! Repeat, using appropriate holes for each size of link. These holes *must* now be at the correct spacing and in line, provided you have held everything firm whilst drilling.

You can now set up the drilled links between centres and turn the outline. Leave the radius on the end till last. I made a little radius form tool for this job. The machined link can be held in the chuck, if you get a piece of thick-walled brass or steel tube $\frac{1}{4}$ in. bore, slit one end about 1 in. long with 3 slits, slide the link inside with the end just sticking out and form the radius. This will leave a very slight parallel portion on the boss, but that won't matter if it blends in nicely. Finally, file the flats; to do this you can again make use of the spacing jig, with two pegs in it, to hold the link in the vice. Leave them a little thick rather than thin—you can adjust when fitting up.

Pivot Bolts, CC & CA (Fig. 29)

These are a bit fiddly, but offer no problems. Machine the $\frac{1}{8}$ in. dia. to be a nice fit in the links you have just drilled.

Drop Arm, CD (Fig. 29)

Drill the holes for this part (2 of them) using the same technique as you did for the spacing jig; the dimensions are important. I made some little buttons the right diameter, with pegs on to fit the holes, to file up the bosses on the end, and then filed the rest of the outline by eye. But you can do it by marking out if you like.

Thrust Links, CE, CF, CG (Fig. 29)

I had no $\frac{5}{16}$ in. × $\frac{1}{4}$ in. stock, so had to mill down a bit of $\frac{5}{16}$ in. square. You need a piece about 3 in. long.

Set this up truly in the 4-jaw, and part off two pieces $\frac{3}{16}$ in. long, first centring each very lightly with your smallest slocumbe drill. Then part off two pieces $1\frac{1}{8}$ in. long, this dimension being fairly important. Remove the pips from the ends, carefully, and then solder the pieces together as shown in fig 33. Use good quality tinmans solder, not the resin core 'wireless' stuff, as by the time you finish there won't be much left of the joint. Now set up the vertical slide and, with a $\frac{3}{16}$ in. endmill, mill out the grooves on each of the $\frac{1}{4}$ in. wide sides. You now have a piece of stock the correct section from which to make two links.

Set between centres, with a small clamp to act as carrier, and machine the profile of the centre shanks, taking care to work from the centre solder joint for longitudinal dimensions. You can make the outline simple, or decorated to taste, but go easy; too much elaboration is not appropriate, and the workpiece is delicate. Now, to drill the holes, use again the 'jig boring' technique with the crosslide, but this time set the zero position to the *centre solder joint* and

Fig 32-Showing the drilling jig in use. The baseblock is the Author's 'universal spacing jig'.

work each way. Don't forget to allow for the backlash in your feedscrews. Centre with a slocumbe, drill No. 14 and follow either with a reamer or a $\frac{3}{16}$ in. drill. Put a little bevel on the edges of the holes, to match the fillet on the pins.

Now you need the pair of semicircular **bushes,** part CF. I make up lengths of split round rod for these; simply saw a piece of, say, $\frac{7}{16}$ in. or $\frac{1}{2}$ in. round rod a few inches long down the middle, file the surfaces smooth, and solder together again. Then centre each end on the joint, and turn it back to a round rod again. This can be used to make all the split bushes you need on the engine, if you start off with the right size, but I keep them in stock, made up as I have time, from about $\frac{1}{4}$ in. to $\frac{7}{8}$ in. O.D. Once you have the split rod, the bush is a straight turning job. Heat to melt the solder and you have 2 off part CF. Before melting apart the thrust links, mark the adjacent parts so that the caps CG go onto the correct ends later.

For the two **straps,** CH, you need some soft material. Blued swedish iron is ideal if you can get any the right thickness (sometimes called planished

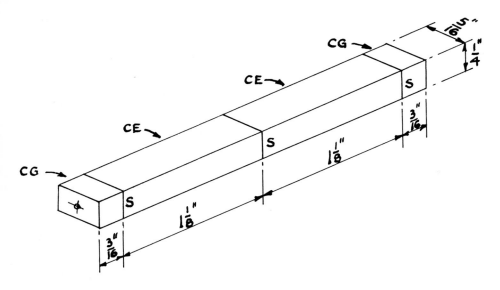

Fig 33-Blanks for the parallel motion thrust links (Part CE etc.).

steel) but if you use bright drawn strip you will have to heat to red and allow to cool very slowly to anneal it before bending. You will need about $3\frac{7}{8}$ in. for each. Bend over a piece of bar just under $\frac{1}{4}$ in. dia. to allow for the springback, and if need be form the backs of the radius with a plastic mallet to take out any bow. Now, having removed the lumps of solder from the faces, assemble the thrust links and their bearing caps inside one of the straps, and clamp up—fingers will do. Mark for the top of the slot in the strap; this mark should be $\frac{1}{32}$ in. *above* the lower edge of the cap CG. You can then mark out the rest of the slot to dimension from this. Drill No. 54 at each end of the slot and open it up with a square swiss file. Finally, file up the ends of the strap $\frac{3}{32}$ in. beyond the bottom of the slot. The little **cotters**, CJ, are a simple filing job. Make the shouldered one first, then the other, and file the latter down till it *just* projects at the small end. Later, when the bearings have bedded in, it will need tapping in further, and you can then cut off any surplus.

Assembly, Beam & Parallel Motion, (Fig. 29, page 52) and (Fig. 34)

You will need a nut to go on the piston rod, and can either make Part DA, or use a normal nut for the present. Press the pins into the beam, using loctite if need be; check that the projection is equal each side of the bosses. If you have used Loctite, leave time for the adhesive to cure. Next, assemble the thrust links CE complete on the beam and the crosshead BF. You will almost certainly have to adjust clearances; if too tight, scrape out metal from the caps—*not* from the body CE; a little scraping here to bed the pin is in order, but any excess removed will alter the working length. If the bearings are slack, file off the shoulders of the caps, again, NOT from the body. When correct, the assembly should just not fall under its own weight. Next, fit the drop arms CD. If the collar nips them too tight, file a little from the face, but better a little tight at this stage than slack. These arms should not be too tight on the pins, but if they are really slack, you will either have to start again, or make little bushes for them. Fit the links BK to the thrust links and then to the drop arms, separately, adjusting as need be, and then assemble between the link and arm. Check that they are still reasonable fits. It should need finger pressure to move them, but no binding. At the other extreme, slack without rattle will do no harm. The cross-shaft BH fits at the joint BK/CD, and not as shown on the drawing supplied by Reeves, which *may* be wrong. (The drawing on page 52 is, however, correct), so this part has to go in when you fit the drop arms to the beam pin. Finally, assemble the short radius links to the drop arms.

Erect the columns and the entablature, taking care that all is square and adjusting if need be. Don't tighten up too hard, as all must come apart again—several times. Use temporary nuts. Attach the beam trunnion bearings to the entablature, and erect the cylinder with piston and rod in place, but don't put in the 'O' rings yet. Drop the beam into its bearings, and check for fit. It should 'just not' fall under its own weight. Bring the crosshead down towards the piston rod, and adjust the cylinder position so that it fits laterally without objecting. Try this at both dead centres, as the cylinder may be slatendicular instead of vertical. If so, you must file the cylinder support to correct it. Alternatively, the beam may not be aligned along the centre of the engine, in which case you may have to draw the holes in the support bearings. In my own case, I made sure that the beam 'looked' right first, because it is a very prominent part of the engine, and misalignment 'by eye' is most obtrusive.

We now have to put in the hole to take the radius link anchor bolt, CA. Make a little scriber out of $\frac{1}{8}$in. silver steel, 60° point, about $\frac{3}{4}$in. long, and harden it. Get a sharp point, but take care that its not off centre. Put the engine on top dead centre, make sure that the piston rod is vertical (cylinder bolts etc. tight, of course) and with the little scriber through the free end of the

Fig 34-The parallel motion assembled on the beam.

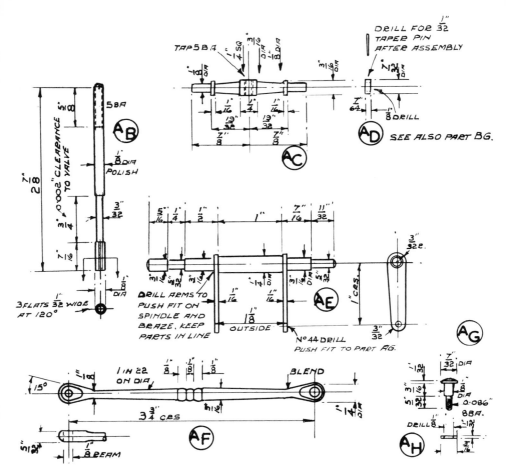

Fig 35-Details of the valve operating gear.

links BJ scribe across the boss of the beam support trunnion. Repeat, with the engine on BDC, *and* with it in mid stroke. In each case make sure that the motion work is in its natural position, not forced in either direction. Don't lean on it, for example!

Take down the beam trunnions, and you will find you have four scribed lines—the one you put in when you first made it, horizontal, and the three you have just scribed. They won't meet exactly at a point, because no parallel motion of the Watt type is truly 'parallel', but they should be pretty close. If they aren't, then something is wrong, and you will have to check all the dimensions and put it right. Now, use an eyeglass and the watchmakers 'foret' referred to earlier (failing this a very sharp and light centre-punch) put a dot in at the mean of the intersections of the lines. Check and recheck with the glass as you enlarge this dot to a centre large enough to start a small drill. I suggest a No. 55. Check that this is true, and enlarge again, finally putting a No. 43 right through. Hold the job in a drilling vice, as it must go through square. Remove the burrs. Now offer up the link with the bolt CA and washer CB, and adjust the latter to give freedom without slackness.

Re-erect the whole issue, with the crosshead free of the piston rod. The gear should move sweetly, if not easily, with no tight places. The hole in the crosshead should drop onto the piston-rod easily—perhaps with a slight scrape—and when you put on the nut you should find the piston slides up and down with just a bit of a scrape here and there. If this scrape is pronounced, or continuous up and down, then adjust the cylinder position for minimum. Finally, put on the 'O' rings, with a few drops of cylinder oil mixed with 3 in 1. You should find all moves pretty well—and looks well, too. You will be able just to detect the movement of the piston rod in the gland due to the inherent fault mentioned earlier (there is only one perfect link type parallel motion, and it was seldom used) but if you have got the dimensions right and assembled as described you will have as good a motion work as will be found on most model engines, and a lot better than on many full-sized ones, too! If you think it looks a bit off, leave it for now and wait till you try it under steam. A little adjustment of the cylinder when hot may correct it.

Valve Rocker etc., Parts AA to AH (Fig. 35)

We might as well finish this part now, and then the greater part of the 'business end' of the engine is done. The **siderods** AF I have already referred to. These are made in exactly the same way as the links in the parallel motion except that they have a bit of decoration on them. Even this is optional. The taper on the shank is about 3° included angle ($1\frac{1}{2}$ setover) or to suit your taste. The **pin** AG should be a good fit in the holes in the links. The little washer may need a bit of adjustment—mine turned out .008 in. thicker than shown on the drawing—to keep the rods straight when erected. The **valve rod** calls for little comment. It is stainless steel, so use a good die for the 5BA thread. The $\frac{3}{4}$ in. long recess for the valve should fit with a few thou. play—it should fit freely without binding when the valve is *hot*. The three flats called for on the end are to let steam or water out of the tailrod cavity in service.

The valve-rod *crosshead*, AC, is made exactly as the engine crosshead, except that it is $\frac{1}{4}$ in. square stock, not rectangular. Again, make the $\frac{1}{8}$ in. ends a good fit in the siderods. You don't want any shake in the valvegear. The **rocker**, AE, is built up. The actual machining and filing calls for little comment except to suggest that you 'jig-drill' the arms to get the centre distances spot on. However, when brazing (or silver soldering) there is a risk of the two arms getting out of line. What I did was to make a $\frac{5}{16}$ in. dia. bush, drilled $\frac{3}{32}$ in. plus, the ends faced truly to 1 in. long, and on assembling this is fitted with a bolt through the two No. 44 holes. Or, you could clamp the two arms to a hefty distance piece, but they might bend when they get hot. However, I have no doubt you will contrive some means of doing the aligning if you don't like my way! You can use any brazing material you like, as the joints can't be seen and the colour doesn't matter.

Assembly of Valve-Motion

Put together the eccentric strap, rod, and the end bearing and adjust approximately to the dimension given on the drawing. The next step will try your patience! You have to wangle part AE, the rocker, inside the cylinder support, and there is only *just* room. If you have made either the oval or the rectangular holes in the support on the small side, it just won't fit. Further, there may be a touch too much thickness in the casting itself inside (casting suppliers are going to reduce this thickness a bit if they can, but there is a limit to how far they can go.) If this is the case, you must either mill or chisel out some from the *inside* front face—the face with the rectangular

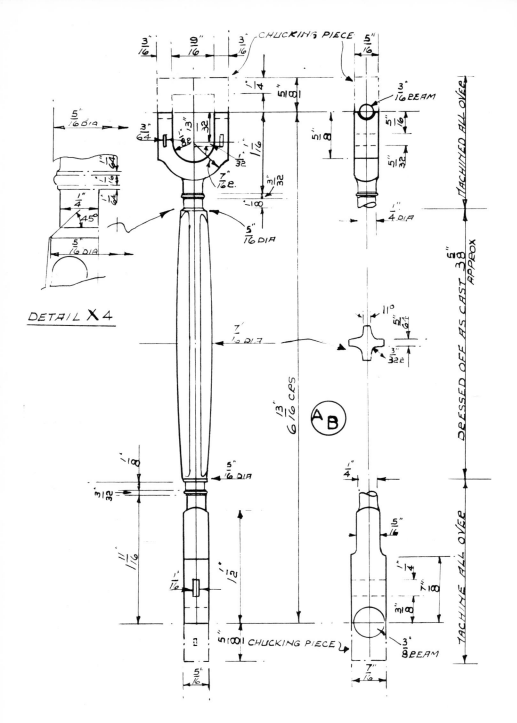

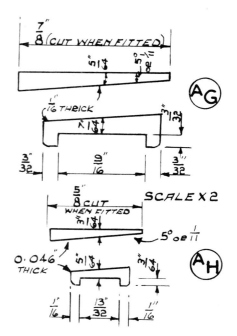

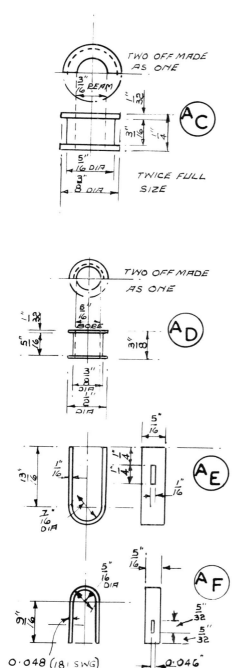

Fig 36 - Details of the connecting rod.

holes. I found it rather like a Chinese puzzle — messed around for quite a time and then — one minute it wouldn't go, the next it was in! However, I should have said, *before* wangling it into place, check that it fits the bearings. Once in place, set up the bearings and fit temporary screws. Assemble the outer lever, but don't pin it; just put a bit of loctite nut lock on it, to give a bit of grip, easily undone. Assemble the eccentric strap to the sheave on the crank, and check that the valve rod end aligns with the outer lever, adjusting the position of the sheave if need be. Fit the crosshead to the valve rod, and then the two side-links can be dropped on. Let these hang straight down, and check the width (thickness) of the little washer AH needed to align them to the rocker arms. In passing, you can adjust the distance apart of these, as they may foul in the cylinder holding down bolts; there isn't much room! Fit the bolts AG, and check that the valve-rod doesn't bottom in the chest. Make sure all is free.

Now you can fit all onto the erecting

base, and put on the flywheel, outer bearing etc., as she is beginning to LOOK like an engine. You can set it up somewhere in the shop whilst you get on with the rest of the work, for it won't need to be disturbed for a while. If you want to make fancy nuts for the column tops, piston and valve-rods, you can carry on. I have made suggestions on the drawings, but you can allow your imagination a reasonably free rein here!

Connecting Rod, Part AB (Fig. 35)

At first sight this looks formidable, but it is, in fact, a relatively easy part to machine if tackled the right way. Trim off the flashes from the casting, and straighten it if need be. In particular, make sure that the rectangular sections of the casting at the two ends are in line. Lay your rule over the job and make a few pencil notes on your drawing if the casting is not quite as drawn. (There is always likely to be a slight difference, as moulds are only made of sand!) These notes should be borne in mind when setting out; for example, it is more important, from an appearance point of view, that the little decorative beads should be equal distance from the cruciform centre than that they should be 'to drawing' from the bearing centres.

Chuck one end in the 4-jaw, and set the nearest part of the cruciform running truly. Take a skim off the cylindrical part adjacent to the shank, say $\frac{7}{16}$ in. wide, so that this can be supported in your fixed steady for the next operation. Reverse, chuck the other end and 'ditto repeato'. Incidentally, whilst doing this job, if the far end of the rod wobbles about a lot, this shows you haven't got it quite straight, and you should correct it, but a small wobble won't matter much. Now, without unchucking, set up the fixed steady on the skimmed part nearest the tailstock. Bring the hard centre up, and adjust the steady till the point, when lightly advanced, does not draw a little circle; this shows that the steady is centralised and you can now put in a fairly deep centre with a slocumbe drill. Reverse and repeat, but make sure that the chuck end is running true. Remove the steady and chuck, and set up the work between centres with the large end at the headstock. Machine the cylindrical sections down to the top diameter of the little bead. Note, by the way, that the bottom end of the rod, carrying a dimension of $1\frac{1}{2}$ in., is rectangular running into a square; the $\frac{5}{16}$ in. dimension is not circular, though it may be cast as such. You will find it will just clean up to square for all that.

To form the bead you can either make a little form tool or turn it as a square-edged shape and then round the top with a fine file; the feature is so small that the latter is probably as good a way as any—it's the way I did it. The 'run in' to the cruciform section should be carried out with some care to avoid cutting into the core of the shank.

Aim for a good appearance; the dimension doesn't matter much. Whilst between centres, take a very light cut at each end and in the centre of the cruciform to act as a guide when subsequently trimming up with a file. Take off just enough to touch each of the ribs, no more.

Now set the centres fairly tight, and with a level or square set the wide part of the fork end horizontal. (fig 37) Mark out the centre-line and the widths of the two ends. Remove from the lathe, and mark out for the centres of the bearing-holes. You may have to effect a compromise, working to the outside of the casting rather than to the drawing dimensions from the decorative beading, but the centre-distance must, of course, be correct. Centre-pop and with dividers scribe a circle the same diameter as the holes are to be. In the case of the top end I marked out on *both* sides, and drilled No. 18 from each, as I felt that the drill might wander through such a deep section, but did not allow the holes to meet in the middle. When drilling these holes use a straight-flute drill, or stone down the cutting edge of a twist drill to

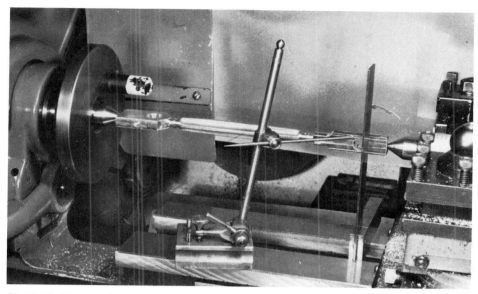

Fig 37 - Marking out the connecting rod between centres in the lathe.

reduce the rake. Examine with a glass when the drill has started and correct the alignment if the point has wandered. You can ream the bottom hole (run the drilling machine at about 500 revs for this) but leave the top for the present, till you have cut out the fork later. You can now either mill or file the sides of the flats to the lines marked out.

Return to the lathe and mark out the centres and widths on the machined surfaces, as before. At the forked end mark out the centres of the two radii and then, with dividers, scribe out the profile of the fork itself. Make a fairly deep centre-pop for the inner radius and put through a $\frac{1}{2}$ in. drill. Be careful; stone the edge of the drill to prevent grabbing, hold the work in a vice, and check that the drill is not wandering. To save subsequent work you can now cut off most of the surplus metal at the ends of the rod but *leave the holes entire* for the present. File or mill the external surfaces and then cut down to the hole in the fork, finally finishing the interior with a file. Aim at a balanced appearance and leave a few thou. on the two $\frac{3}{16}$ in.

wide arms. You can now poke a $\frac{3}{16}$ in. reamer through the top end holes, or, if you haven't got one, increase the size of the hole by one drill number at a time till you reach No. 13 and then follow with $\frac{3}{16}$ in. You can now cut both ends through the centres of the holes, leaving a little on to be filed off when fitting the bearings.

Bearing Bushes, Parts AC and AD (Fig. 36 page 63)

These are made in exactly the same way. I start by soldering together *flat* brass strip to give me a piece of square stock which can be turned down. This is easier and, in the event, more economical than my usual method sawing a piece of round stock down the middle; the latter, when soldered together, has to be held in the 4-jaw anyway. The parts are normal turning jobs and need little comment. Make the bottom end bush, AD, a snug fit to the end of the rod. The outside width of the bushes should, of course, be made to fit the crankpin or beam pin as appropriate. When unsoldered, clean off the solder but don't go right down to the brass; leave a tinned appearance. You have already made some straps and cotters for the parallel motion, and those for the rod are dealt with in the same way. One detail is missing from the drawing; you need a little packing piece to go between the bottom half bush AD and the strap AE. This is made from softened copper or brass strip $\frac{1}{32}$ in. thick. As before, make the wedges well over length, and cut to size after all is erected.

Preliminary Test

You now have enough parts made to run the engine. Assemble the connecting rod to the beam and check that the bottom end falls in line with the crank. If it doesn't, you will have to adjust either the position of the main bearings or the beam trunnions. But make sure that the error is not due to the holes in the fork being out of line; if they are, then you may have to bend the rod a trifle. Note that there should be only a running side clearance on the top end, but a little extra can be allowed on the crankpin. The fit of the bearings at this stage should be such that the rod only just falls under its own weight if held horizontal. Once this part is in order you can check that the engine can rotate with reasonably equal clearance at both ends of the cylinder, and that there is no binding anywhere.

You will have to dismantle the cylinder and valve chest to make paper joints if not already done. Oiled brown paper is satisfactory, but I use thin 'Oakenstrong' jointing, obtainable from Stuart Turner Ltd. This *is* strong, as its name suggests, and is easy to cut. I mark out holes for studs etc. with a sharp point, and then punch them out with a leather punch. Reassemble, and adjust the position of the valve so that it does not foul either end of the chest when moved through its full travel from the eccentric. Don't fit the steam-chest cover—put on a couple of nuts with washers below to hold the chest in place. Oh—I nearly forgot! If you are fitting a drain-cock as indicated on the cylinder drawing this must go on before attaching the valve chest as there isn't enough room to turn it otherwise.

Valve Setting

Put the engine somewhere where there is good light on the steam-chest, and set the piston on dead centre. *Note*, this may not be with the *crank* on dead centre, owing to the geometry of the gear. Adjust the eccentric till the steam port is just opening, rotating the eccentric the right way. (Beam engines normally run with the connecting rod moving *away* from the engine in the upper half of the crank circle). Tighten one eccentric grubscrew finger tight, and turn the engine to the opposite dead centre, again make sure it turns the correct way. If you are lucky, the valve will be just opening to steam at this end; if not, adjust the valve-rod in its crosshead to move the valve as necessary. Repeat this procedure till you get equal openings at each dead centre. Make a final adjustment of the eccentric to bring this opening to not less than .002 in., with feelers, and don't make it more than .005 in.

Note that the rocking lever (AA on drawing No. 4) is only fixed with loctite at this stage; leave it as it is, but when you come to final erection see that this lever stands vertical with the valve in mid-travel, and then fit the proper pin.

Make a little stub of pipe to push into the hole in the steam chest cover, and, if you have already drilled the holes for the inlet flange, put in some small screws to block them for the present. Squirt a little oil (something like SAE 30 motor oil will do, or Vitres 27 as used on the lathe) into the ports and fit the steam chest cover. Now, if you apply a few pounds/sq.in. air pressure she should turn over quite happily. She may be a little tight, but so long as there are no grunts and groans at particular points in the stroke, not to worry. A 'wheeze' from the cylinder may be no more than the 'O' ring running over the invisible ridges left by the boring tool, but if it happens only at one or other end of the stroke, slightly slacken the four cylinder holding-down bolts and adjust the cylinder position for 'least noise'. A similar grunt at the crank end calls for adjustment of the main bearing positions.

If she grunts at *both* ends at the same point in the stroke, the engine is telling you that the beam is out of alignment and you should try adjusting the beam trunnions. There is no mystery about this stage; the engine knows very well what, if anything, is wrong, and all you have to do is to study her language and do what she says. (Babies are far less intelligible, but Mother soon learns what they mean!)

As soon as the engine sounds happy, if stiff, let her have enough air to run at about 60 rpm, and run for a while. If the speed increases, all is well, and the bearings are bedding down; if the reverse, look for the tight spot and get busy with a scraper. After about half-an-hour's run, take down the connecting rod and crank bearings and look for the bedding marks, and touch these with a scraper if you think they need it. (Usually you will find hard binding at the fillets in the corners.) On the other hand, if there is a slack bearing with a knock, take off a very little metal from the bearing joint faces and re-assemble. Don't expect to achieve an easy, but well fitted bearing; this will take many hours of running. The aim is to remove all tight spots, leaving some resistance to motion but a velvety feel about it. A final point to look at is the bedding of the slide valve. In this case the oil does most of the sealing—the engine will never need more than 5lb.sq.in. at the chest to keep it going, and when well bedded will turn over on much less—but any signs of uneven bearing should be corrected.

You can now reconnect the air supply, send for the family, and stand back whilst they tell you how clever you are!

Pump, Parts G & H (Dwg. 3), and DB/DD (Dwg. 4) (Fig. 38 and Fig. 30, p. 55)

I seem to have forgotten this bit, perhaps because there is so little work to it. The pump body casting has the gland cast as one with it, and the first job is to saw this off and trim up the casting. Set

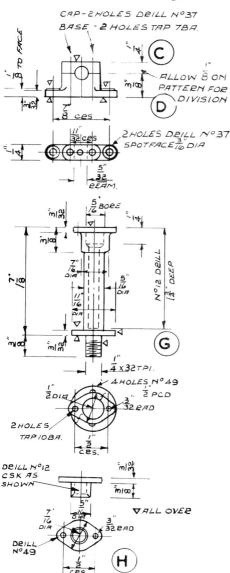

Fig 38-Detail of pump body.

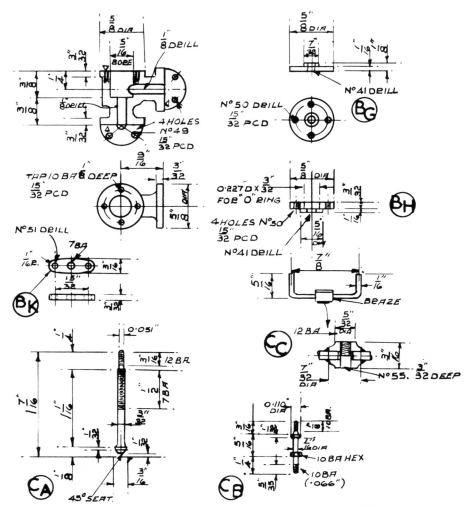

Fig 39-Details of stop valve. Note—the 0.227 in. dimension shown on part BH may be gauged from the shank of a No. 1 drill.

true in the 4-jaw by the base and centre drill the other end. Support this end in the tailstock and machine the outside of the body, including the lower flange. Face the top end. Reverse and grip lightly so as not to mark the work to face the bottom flange and form the screwed spigot. (This can be sawn off if you are going to use the pump only as a dummy, as intended; but if you want it to work you must make up a little valve-box to screw on.) Now grip in the 3-jaw either by the flange or the spigot and put a $\frac{5}{16}$ in. drill down $\frac{1}{4}$ in. deep, followed by No. 22 to $1\frac{1}{4}$ in. deep, and finally a small hole, say No. 35 at the bottom. Machine up the gland as described for those on the valve-chest, fit this into the top of the barrel and file up the ovals to match. Finally, mark out for and drill the 4 holding down holes and the two stud-holes for the gland.

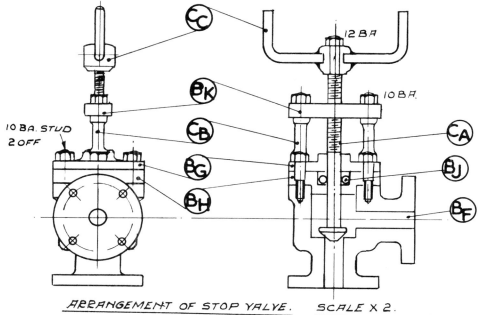

Fig 39a ARRANGEMENT OF STOP VALVE. SCALE × 2.

The other bits and pieces on fig. 30 require little comment. Make the pump-rod, DD, last, after you have measured the distance between the parallel motion cross-bar and the pump body, to make sure that the ram neither bottoms nor comes out of the gland. The latter is packed with a little oily gland-packing, or even cotton string, not too tight; just enough to stop the ram from rattling about.

Stop-Valve (Fig. 39)

The casting has two chucking pieces and these should be trimmed with the file as well as the main body. Chuck by the base piece and set true. To machine the $\frac{5}{8}$ in. dia. top you will have to set up a narrow tool — a parting tool, not too deep in section will do — and feed in axially, as the side flange prevents normal approach. Face the end, centre, and drill and bore to $\frac{5}{16}$ in. × $\frac{3}{8}$ in. deep. With a small slocumbe, centre the base of the cavity and drill No. 31 followed by either a $\frac{1}{8}$ in. reamer or a drill. Now chuck by the other chucking piece and machine the side flange; you will have to use a parting tool to face the back of the flange and reduce the diameter of the body behind the flange to $\frac{5}{16}$ in. dia. to clear the nuts. Drill $\frac{1}{8}$ in. to break into the central cavity.

Make a stub-mandrel to hold the casting whilst machining the base flange, after sawing off the chucking-piece. Again, you will need narrow tools to get behind the flange. Finally, saw off the side piece and file up the rest of the casting. Don't mark out for the stud-holes etc. — this is best done from the mating parts.

Most of the rest of the bits are straightforward but you may have difficulty in holding — or even, perhaps, seeing! — the two pillars CB. If this is so, make up a couple of studs from $\frac{1}{16}$ in. material, with a plain section in the middle, and screw on two 10BA nuts. I know the top collar *ought* to be round, but I doubt if you will be criticised (except by yourself) if they are hexagon. Unless you have a watchmakers lathe you will have to machine CA in stages, with $\frac{1}{4}$ in. sticking

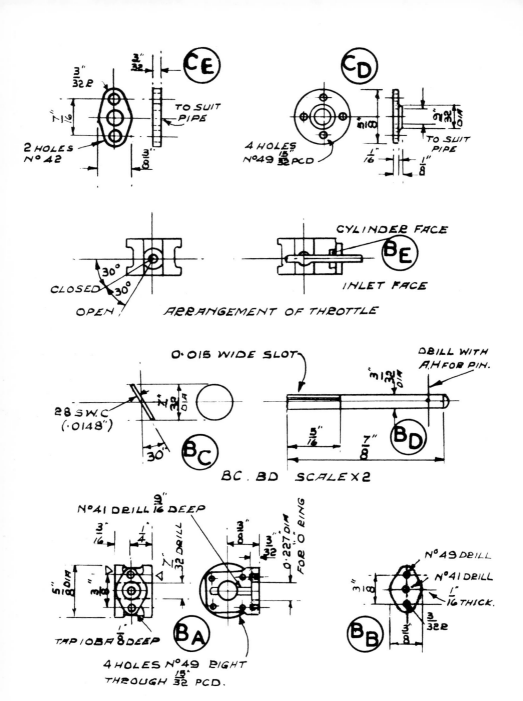

Fig 40-Details of the governor throttle valve. See text when machining part BC.

out of the chuck at a time. If you run into trouble, make the part from a piece of $\frac{3}{32}$ in. stainless and screw on the valve seat with a little loctite to hold it there. We are not dealing with high-pressure super-heated steam! If you haven't a 12BA tap and die, make this part and CC 10BA. The tip with the handle is to cross-drill first, then chuck the stock, machine the profile, drill and tap. The cross-drill should be a tight fit on the $\frac{1}{16}$ in. silver steel used for the handles. I expected trouble brazing this up, but with Easiflow silver solder and flux nothing slipped. The ends of the handles were trimmed with a file afterwards. Mark out and drill the flange BG and then use this to drill the mating parts, but you will need the governor casing to deal with the side-flange. Incidentally, it will be easier if you make the holes in the bottom flange of the valve tapped 10BA instead of the No.49 clearance hole, and attach the steam inlet flange with set-screws. On my own engine I also tapped the inlet passage of the valve $\frac{5}{32}$ in. × 40, so that for test purposes a pipe could be screwed in direct, with Part CD only a decoration!

Governor Valve, Parts BA to BD (Fig. 40)

The body presents only one difficulty—to see that the cross-drill is on the same centre-line as the bore. This can only be ensured by careful marking out and drilling. A *small* error will not matter, but the dimensions themselves are small, and you must work to the closest possible limits. The .227 in. bore need not be flat bottomed as shown, and I found the shank of a No. 1 drill made a good gauge; the hole should be a trifle small rather than large.

To make the disc, BC, file off the end of a piece of $\frac{1}{4}$ in. dia. brass at 30° and solder a small piece of 28 gauge brass to this. Chuck in the 3-jaw, and machine both stock and the sheet to an easy push fit into the bore of the body; this ensures a correct shape and profile. Remove all traces of solder after detaching the disc. The $\frac{1}{64}$ in. wide slot in the spindle may present some difficulty if you haven't got a slitting saw of this size. They are obtainable relatively cheaply from various sources but if you can't get one the only alternative is to grind down an Eclipse mini-hacksaw blade.

It's a bit of a fiddle getting the spindle into the body and engaged with the disc, and unless you are careful you will find that it's the wrong way round when you *do* succeed. The drawing does, however, show quite clearly the relative positions of the links and the disc. Once in place poke a $\frac{1}{32}$ in. drill through spindle and disc together, fit a short length of copper wire, and when you are satisfied that the spindle still turns easily, put on a dab of solder. If you haven't a soldering iron that will reach, put on a dab of Araldite instead, and leave the assembly in a hot place to cure.

There is sometimes one problem with a governor valve of this description; the slit ends of the spindle may open out after soldering—possibly due to solder travelling down the slit by capillary action. I suggest that, though it is not shown on the drawing, you turn the end of the spindle about 0.006 in. small the full length of the split part. Incidentally, the thickness of the valve disc BC is by no means critical; I have used 0.006 in. shim brass before now. The disc is so well supported that it will be quite strong enough.

Having marked out and drilled the various holes (there is no harm in making the four fixing holes a couple of drill numbers larger than shown to ease assembly if desired) put a drop of oil on the spindle, fit the 'O' ring, and bolt on the oval flange. The spindle should turn quite easily. Don't worry too much if the valve jams slightly when closed as you can set up the governor gear so that it stops before this point is reached.

Governor (Fig. 41)

Have a good look at the general arrangement, and make sure you understand how the thing works, and especially which way

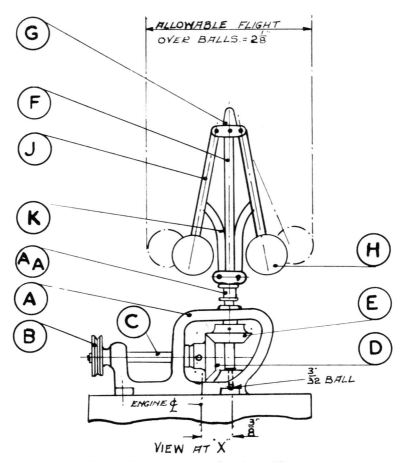

Fig 41 -Arrangement and layout of governor gear. (See also p. 73).

the rods move when the speed rises (i.e. when the balls fly out) Don't be alarmed at all the pin-joints. These present few problems, especially if you use Loctite nut-lock (but *not* the stronger varieties) to make temporary joints whilst drilling. The governor does work—or, at least, serves to limit the maximum speed—but will only do so if the whole of the link-work is a very easy, but not sloppy, working fit. For this reason I suggest that you drill all holes except those for the $\frac{1}{32}$ in. taper pins one size smaller than shown, and enlarge them by one drill number at a time till you get the easy fit.

You must also be careful to remove all burrs, and ensure that working surfaces are polished. I would also advise that you don't locate the part AF (the pillar) till you have made all the rest, as its position will depend on the cumulative errors that may arise in manufacture.

Pendulum and Drive, Parts A to AA (Fig. 42)

('Pendulum' is the proper name for the rotating ball assembly!) Most of this is straightforward turning and file-work, needing no special procedure but pos-

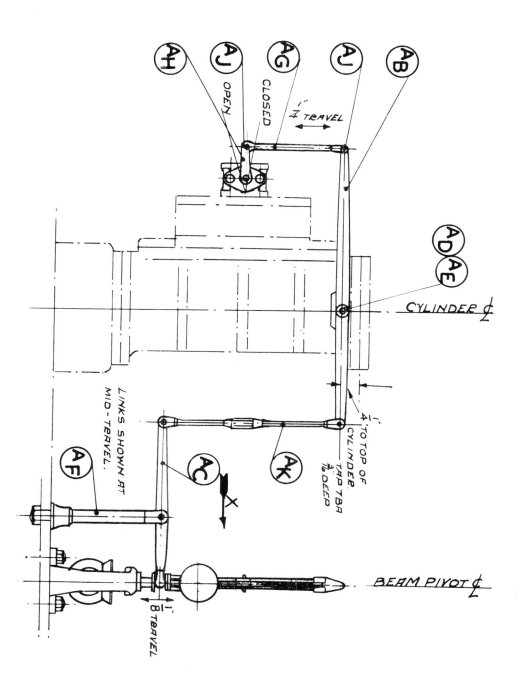

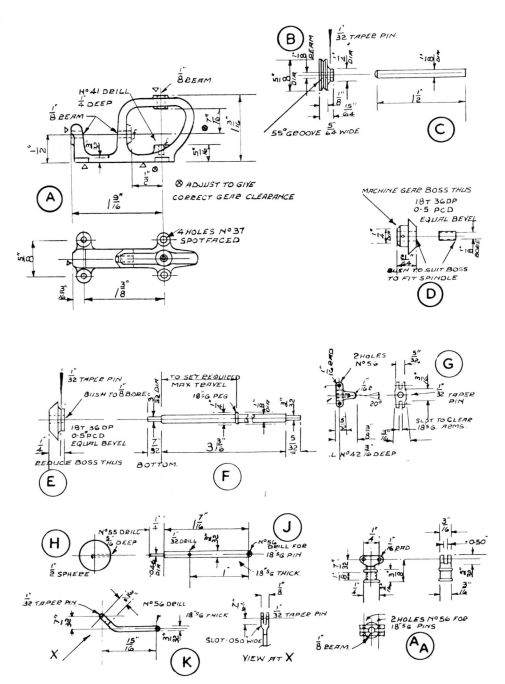

Fig 42-Details of gear bracket and governor parts.

sibly a suitable vocabulary when it comes to holding the smaller bits. For these I use a small hand vice or the Eclipse 180 instrument vice, held in the jaws of the bench-vice. At the time of writing (a good many months before you read it) it seems possible that the gears may be supplied modified to drawing, and you will be saved that chore. If, however, you *do* find you have to modify them, I recommend that you bush them first, and then reduce the bosses to fit the space in the bracket.

The **bracket**, 'A', should first be filed on the underside till it sits flat and with the body vertical. Clean up the top and end faces, and then mark out, taking great care to get the vertical and horizontal holes in line. Drill the outer holes No. 31, follow with $\frac{1}{8}$ in. reamer or drill, and then make a little punch out of $\frac{1}{8}$ in. silver steel and centrepop the inner face; don't rely on the drill following straight from the outer hole — it never does! Drill this bottom hole (vertical) No. 41, and the inner (horizontal) one 31 and then ream both. Mark out for, drill, and spotface the four fixing holes. Don't do any work on the bosses until you fit the gears, but clean up the outside of the casting.

The **Spindle** 'F' needs care. Check that the piece of stock is dead straight by rolling it on the lathe bed. Unless you have accurate collets I recommend that you turn down the ends in the 4-jaw, using a bit of paper under the jaws to avoid marking the shank. Aim for a good finish on the bottom peg, as this is a bearing. The **weight-carrier** 'G' is made from $\frac{1}{2}$ in. $\times \frac{3}{16}$ in. stock held in the 4-jaw. Drill the hole from the tailstock after facing the end and then use your surface gauge or scribing block to mark out on the end for the slot and on the side for the $\frac{5}{16}$ in. centre-distance of the cross-holes. Now with a narrow slightly round-nose tool form the profile of the 'steeple'. The angle doesn't matter a bit — it only has to look well. When nearly to shape, part off — and take care you don't lose it in the swarf on the lathe-bench!

Make a little stub mandrel, mount the thing on this in the No. 42 hole, and finish the profile. Incidentally, if you have made the stub too slack, put a little nut-lock Loctite on, and do something else whilst it cures; this will grip well enough if you take light cuts. Drill the cross-holes and then cut the slots. If you haven't a slitting saw the right thickness, — or if you can't see how to hold it — you will find an ordinary 24 tooth hacksaw just undersize, and the slot can be widened with a thin warding file. Part 'AA' is made in a similar fashion, but of $\frac{1}{4}$ in. $\times \frac{3}{8}$ in. this time. Incidentally, both of these can be of steel if you like. In this case it is most important that the $\frac{1}{8}$ in. reamed hole is an easy slide fit on the spindle 'F' and if you can't achieve this with your reamer, try a 3.2mm drill, or No. 30 if you have none metric. Note that the slot goes right across.

The **links** 'J' and 'K' should be made or reasonably stiff material — bright drawn is O.K., but if your supply of material is annealed or soft you may have to hammer it a bit to harden it. It is important that the holes be at identical distance apart and that the $1\frac{7}{16}$ in. dimension is the same on both of part 'J', so file and drill them as a pair. You can make the forked end link 'J' either by filing up a piece of $\frac{1}{8}$ in. square stuff or, as I did, by brazing on the cheeks of the fork. I made these cheeks well oversize and filed them down afterwards; any surplus silver solder was also removed.

There is a trick you can use to get the centre-distance of the holes the same. Drill the forked ends of both, with a piece of material between, and remove the burrs. Bend the arm to the necessary shape, getting them the same by eye. Now drill *one* arm at the other end, working carefully to the $\frac{15}{16}$ in. and $\frac{7}{32}$ in. figures. Slip the plain end of one into the fork of the other and put a pin through the holes; now you can drill the other plain end through the hole in the fork, and be sure that both are alike. (Its more important that they be the same than exact to dimension.)

Incidentally, I should have said that

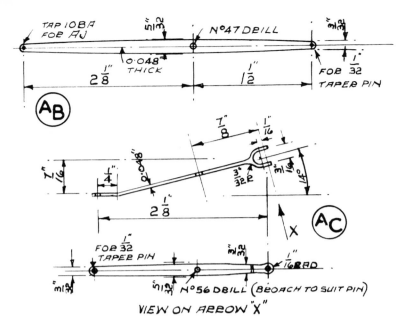

Fig 43-Details of linkwork between governor and throttle.

the .046 in. dia. on the end of 'J' can be square. Mike across the corners, and drill the balls one drill size smaller than this figure. Simply hold the ball in the 3-jaw, centre, and drill. The ball is attached to the arm with Araldite. Don't use Loctite for this job; araldite is an adhesive that cures by the action of a hardener additive, but Loctite cures only in the absence of air. It may *not* cure if air can get down the hole. The other bits, like the pulley and so on, need no comment. Once you have made these you can put the assembly together.

Take the two gears—look at fig 48 again—and offer them up in the bracket. You will soon see whether you need to take anything off either boss. Bring this nearly right, and then again offer up the gears, with the spindles in this time. Keep on reducing the bosses till the gears mesh nice and freely. A little backlash should be allowed. If you see that you may have to reduce the bosses more than they will stand, take a bit more off the gear bosses. If you take off too much in error, a washer is in order, but make sure that it is not obstrusive.

The assembly of the arms to the carriers is fiddly. I suggest you attach the top hanger to the spindle temporarily with Loctite for this job. Use undersize pins to start with, then when you are sure all is in order, replace them one at a time with the proper sizes, broaching the holes if need be. Set the assembly in place on the engine bed, and note how far the sliding sleeve moves when the weights just touch the engine columns. Drill for and fit the little pin in the spindle just below this point, so that the balls can never hit the columns. I recommend that you leave all as it is till it has been tested, but once that is done, you can rivet over the ends of all pins after cutting off any surplus.

Governor Linkwork, Parts AB to AK (Fig. 43)

I need say very little about the parts on this section of the drawing, as about the only components that will cause any trouble are the levers 'AC' and 'AH'.

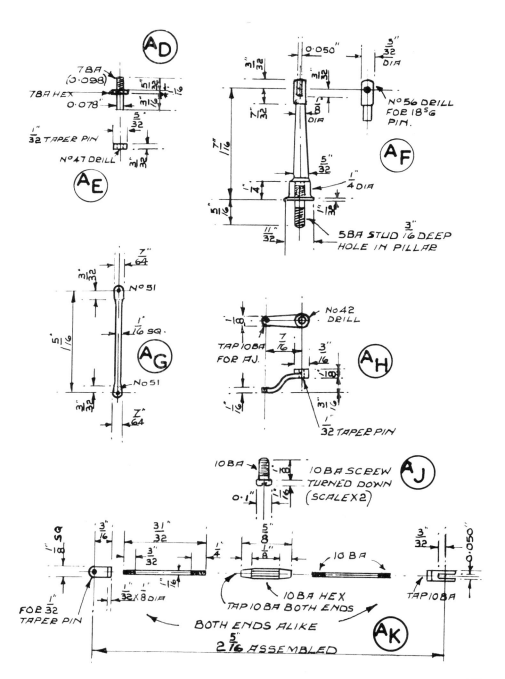

(Trouble other than holding the little bits, that is!) The rocking lever, 'AH' is one part of the design of which I took from the model beam engine to which I referred in the first article. That one, believe it or not, had been forged in one piece. If you want to do it that way, take a piece of stock and bump up a knob on one end, then flatten the main part of the arm to the required thickness. Now attack the 'bump' on the end, and forge two flat ears on either side. File these roughly to shape, and then forge round a mandrel, finally forming the shape with a file. Easy, if you have a three-inch anvil, a small enough hammer, and can keep the job hot whilst you do it! The better way is to make the fork separately, and then braze it to the rest. You will still have to file up the little rounded ends—these should be an easy fit in the $\frac{1}{8}$in. wide slot in part 'AA'.

Don't put the bend in the other end till you have offered all up to the engine; it may need to be slightly different from that shown.

The other lever, 'AH' is again best made in two pieces silver-soldered together. Make a little boss with a shoulder on it to press into a $\frac{1}{8}$in. hole in one end of the unbent arm; You can drill the other hole in the arm now, too, if you like. The centre distance before bending is almost exactly $\frac{1}{2}$in. so make it so. Press in the boss after bending the arm, and braze up. I don't think the other parts will cause you any difficulty, but you can make 'AJ' as a hexagon screw with advantage to appearance and the pillar 'AF' is better made of steel if you are not going to paint it. Don't make the adjusting lik till you have offered up the rest of the gear, when you can measure the correct length (shown $2\frac{5}{16}$in. on the drawing).

Fig 44-'MARY' as built by the Author and shown at the Model Engineer Exhibition. This version has fluted columns and much refined detail. All steel parts are stainless.

Assembly

Mark out for and drill the holding down bolt-holes for the pendulum bracket and fit this in place. Set part AB in place on the cylinder and wedge it horizontal. With the stud removed, stand the pillar AF, with AC in place, on the bed and engage the forked end with the sleeve AA. Adjust its position till the end of the lever lies underneath the lever AB; bend AC to line up. Mark the position for the pillar, and drill the fixing hole, after which procedure you can attach this to the bed.

Attach the lever AH to the spindle of the governor valve, using Loctite nutlock for the present, bearing in mind the position shown on fig 48, and when it has cured fit up the rod AG. Wedge up this assembly so that the governor valve is closed, and also wedge up the sleeve AA so that it is about .020 in. below the travel limiting pin. You can now measure the distance between the holes in AB and AC and make AK to suit, leaving a little for adjustment both ways. Once you have this part, you can fit it, and make a fine adjustment with the screws to allow for the collective slack in the pins; the valve should close when the sleeve is .020 in. below the stop-pin. Under these conditions, if the engine runs away the governor will close the valve and limit the top speed — in my case, to 140 rpm with air at 10 lb.sq.in. (It needs only 1 lb.sq.in. to run normally, by the way.)

The major problem with model governors of this size, apart from that of 'scale speed', is that though there is enough force to *close* the valve, the balls are not heavy enough to open it again smartly. If this is the case, don't add springs or weights, as this will act as a speed increasing gear. The only remedy, after going over the gear to remove any friction or stickiness, is to substitute lead balls for the bronze ones. (This will have *no* effect on the equilibrium speed of the governor.) These can be had in various sizes from gunsmiths, as they are used by the happy band of enthusiasts who preserve (and fire) antique muskets.

Conclusion

I have said nothing about painting the model, as this is very much a matter of personal taste. Green is a popular colour with details picked out in black or red — or even gold — but if you *do* use green avoid the bilious grass-green found on so many models. The 'British Racing Green' used on some Austin and Morris cars is tolerable. A very popular colour in the 19th century was Brown — very similar to that used on the old 'Bull Nosed Morris' of the '30s. A suitable alternative may be found from the Porsche range of a few years ago. It is most important to use very thin coats of paint, as the type of enamel used on model locos. for example, is so thick as to obscure the fine detail. I use cellulose primer and for general painting cellulose colour, applied either from the well-known Aerosol can or an air-brush; the latter being useful for the smaller parts.

To do the painting properly, all will have to be dismantled, and you should take the opportunity of dealing with any evidence of tight spots — or the converse, if knocks have appeared during the running trials. Before doing so, however, drill and tap for 12 or 10BA round-head screws to attach the nameplates on the sides of the entablature. Messrs Reeves tell me that they will happily supply other names if 'MARY' might lead to domestic strife, but that *is* her proper name!

So, we have come to the end. I think you will agree that you now have a really elegant example of the steam age, and a model you can be proud of. I certainly enjoyed making mine, and hope that you will have as much fun with yours.

Schedule of parts on castings suppliers drawing Nos. RV2, RV3, RV4, RV5

'STANDOUT' applied to studs is the length projecting when screwed into the casting.

DWG. RV32 Sheet 2

Part Letter	Description	Remarks	Material Supplied	Qty.
A	Cylinder Barrel		Gun Metal	1
B	Steam Port Cover		Gun Metal	1
C	Valve		Brass	1
D	Steam Chest		Gun Metal	1
E	Chest Cover		Gun Metal	1
F	Piston	$\frac{7}{8} \times \frac{3}{32}$ in. 'O' Ring	Gun Metal	1
G	Top Cylinder Cover		Gun Metal	1
H	Bottom Cylinder Cover		Gun Metal	1
J	Piston Rod Gland	$\frac{5}{32}$ in. 'O' Ring BS 007	Gun Metal	1
K	Valve Rod Gland	$\frac{1}{8}$ in. 'O' Ring BS 006	Gun Metal	1
AA	Cylinder Pedestal		Gun Metal	1
AB	Connecting Rod		Gun Metal	1
AC	Top End Bearing Shell		Brass Bar	2
AD	Bottom End Bearing Shell		Brass Bar	2
AE	Bottom End Strap		Mild Steel	1
AF	Top End Strap		Mild Steel	2
AG	Bottom End Cotter		Mild Steel	1 set
AH	Top End Cotter	1 Set Identical Required for Eccentric Rod—Sheet 4 item c	Mild Steel	2 sets
AJ	Beam		Cast Iron	1
AK	Beam Trunnion	Cap Cast with Standard	Gun Metal	2
BA	Piston Rod	Rustless Steel	EN38	1
	Nuts 8BA	(4 are Locknuts)	Mild Steel	10
	Nuts 7BA	(2 are Locknuts)	Mild Steel	28
AK	Bearing Cap Stud	7BA $\frac{3}{8}$ in. Standout	Mild Steel	4
HI	Bottom Cover Screws	7BA Csk. Head $\frac{1}{4}$ in. overall	Mild Steel	8
GI	Top Cover Gland Stud	8BA $\frac{3}{8}$ in. Standout—Nut and Locknut	Mild Steel	4
DI	Valve Gland Stud	7BA $\frac{3}{8}$ in. Standout—Nut and Lockout	Mild Steel	2
A4	Exhaust Flange Stud	8BA to suit Pipe Flange	Mild Steel	2
A3	Steam Chest Stud	7BA $\frac{3}{4}$ in. Standout	Mild Steel	8
A2	Part Cover Stud	7BA $\frac{1}{4}$ in. Standout	Mild Steel	4
A1	Top Cover Stud	7BA $\frac{1}{4}$ in. Standout from Casting	Mild Steel	8

DWG. RV32 Sheet 3

A	Bedplate		Gun Metal	1
B	Outer Bearing Support		Gun Metal	1
C	Rocker Bearing Cap	Cast as one with Part D	Gun Metal	2
D	Rocker Bearing	See Part C	Gun Metal	2
E	Flywheel		Cast Iron	1
F	Entablature		Gun Metal	1
G	Pump Body		Gun Metal	1
H	Pump Gland		Gun Metal	1
J	Inner Main Bearing Cap	Cast as one with K $\frac{3}{8}$ in. bore	Gun Metal	1
K	Inner Main Bearing	See Part J	Gun Metal	1
AJ	Outer Main Bearing Cap	As Part J but $\frac{5}{16}$ in. bore see note	Gun Metal	1
AK	Outer Main Bearing	As Part K but $\frac{5}{16}$ in. bore see note	Gun Metal	1

Part Letter	Description	Remarks	Material Supplied	Qty.
F2	Nameplate Fixing Screw	To suit nameplate	Brass	4
	Locknut 10BA	For GI	Mild Steel	2
	Nut 10BA		Mild Steel	6
	Nut 7BA		Mild Steel	10
	Locknut 5BA	For JI	Mild Steel	4
	Nut 5BA		Mild Steel	16
JI	Bearing Cap Stud	5BA $\frac{5}{8}$ in. Standout (2 each J and AJ)	Mild Steel	4
GI	Feedpump Gland Stud	10BA $\frac{5}{16}$ in. Standout	Mild Steel	2
FI	Pivot Bearing Stud	5BA $\frac{5}{16}$ in. Standout	Mild Steel	4
DI	Rocker Bearing Cap Stud	7BA $\frac{3}{8}$ in. Standout	Mild Steel	2
BI	Main Bearing Stud	As AI	Mild Steel	2
A5	Rocker Bearing Stud	7BA $\frac{1}{4}$ in. Standout shorten as req.	Mild Steel	4
A4	Cylinder Stud	5BA $1\frac{1}{16}$ in. Standout	Mild Steel	4
A3	Feedpump Stud	10BA $\frac{5}{32}$ in. Standout	Mild Steel	4
A2	Governor Bracket Stud	7BA $\frac{1}{4}$ in. Standout	Mild Steel	4
A1	Main Bearing Stud	5BA $\frac{5}{16}$ in. Standout	Mild Steel	2

DWG. RV32 Sheet 4

Part Letter	Description	Remarks	Material Supplied	Qty.
A	Eccentric Strap	Casting supplied in one piece	Gun Metal	1
B	Eccentric Sheave		Cast Iron	1
C	Eccentric Bolt		Mild Steel	2
D	Eccentric Rod		Mild Steel	1
E	Eccentric Rod End		Mild Steel	1
F	Bush Rod End	Drawn Rod	Gun Metal	1
G	Bush Loose Flange		Brass	1
H	Bearing Strap		Mild Steel	1
J	Strap Cotter	Use item AH Dwg Sheet 2	Mild Steel	1 set
K	Lever Pivot Bolt		Mild Steel	1
AA	Rocking Lever		Mild Steel	1
AB	Valve Rod	Rustless Steel	EN58	1
AC	Valve Crosshead		Mild Steel	1
AD	Retaining Collar	Identical to BG	Mild Steel	2
AE	Rocker Shaft		Mild Steel	1
AF	Side Rod	Drilling Jig same as for BJ, BK	Mild Steel	2
AG	Rocker Pivot Bolt		Mild Steel	2
AH	Pivot Bolt Sparing Washer		Mild Steel	2
AJ	Crankshaft	Built-up	Mild Steel	1
AK	Flywheel Key		Mild Steel	1
BA	Column	Made in parts	60/40 Brass	4
BB	Beam End Pivot Pins		Mild Steel	2
BC	Main Beam Pivot		Mild Steel	1
BD	Intermediate Beam Pivot Pin		Mild Steel	1
BE	Retaining Collar	For $\frac{9}{64}$ in. pin	Mild Steel	2
BF	Piston Rod Crosshead		Mild Steel	1
BG	Retaining Collar	Identical to AD	Mild Steel	4
BH	Parallel Motion Cross-Shaft		Mild Steel	1
BJ	Radius Link	Drilling Jig as for AF and BK	Mild Steel	2
BK	Parallel Link	Drilling Jig as for AF and BJ	Mild Steel	2
CA	Radius Link Anchor Pivot	Attached to Dwg. Sh2/AK	Mild Steel	2

Part Letter	Description	Remarks	Material Supplied	Qty.
CB	Pivot Spacer		Mild Steel	2
CC	Drop Arm Pivot Bolt		Mild Steel	2
CD	Drop Arm		Mild Steel	2
CE	Thrust Link Body		Brass	2
CF	Thrust Link Top Bearing	Turn as a pair	Brass	2
CG	Thrust Link Bottom Bearing		Brass	2
CH	Thrust Link Strap		Mild Steel	2
CJ	Strap Cotter		Mild Steel	2 sets
CK	Valve Rod Locknut		Mild Steel	1
DA	Piston Rod Locknut		Mild Steel	1
DB	Pump Beam		Brass	1
DC	Pump End Rod		Mild Steel	1
DD	Pump Rod		Silver Steel	1
DE	Column Top Stud		Silver Steel	4
DF	Column Lower Stud		Silver Steel	4
DG	Column Top Nut		Mild Steel	4
DH	Dummy Lubricator	For Main Bearing and Beam Trunnions	Brass	4
DJ	Nameplate	Cast or Photo-Engraved 18's Gauge	Yellow Brass	2
	8BA Locknut		Mild Steel	2
	8BA Nut		Mild Steel	9
C	7BA Nut		Mild Steel	2
DI	5BA Locknut		Mild Steel	2
BI	6BA Setscrew	$\frac{1}{8}$ in. A/F Sq. Head Thread $\frac{3}{16}$ in. Shank	Mild Steel	2
F2	3BA Turned Washer		Mild Steel	4
F1	3BA Std. Nut		Mild Steel	4
JI	R/Hd. Screw	10 BA × $\frac{1}{8}$ in.	Brass	2
CI	$\frac{1}{16}$ in. Parallel Pin	For DC	Silver Steel	1
AI	$\frac{3}{64}$ in. Taper Pin	For AA A. G. Thomas	Mild Steel	1
	$\frac{1}{32}$ in. Taper Pin	A. C. Thomas, Heaton Road, Bradford	Mild Steel	8

DWG. RV32 Sheet 5

A	Governor Bracket		Gun Metal	1
B	Pulley		Brass	1
C	Driving Spindle		Mild Steel	1
D	Driving Bevel 18T 36DP $\frac{1}{2}$ in. PCD	Bush and Modify as shown. 6BA Grub Screw	Brass	1
E	Driving Bevel 18T 36 DP $\frac{1}{2}$ in. PCD	Bush and Modify as shown, for Taper Pin	Brass	1
F	Governor Spindle		Silver Steel	1
G	Weight Hanger		Brass	1
H	Weight	$\frac{1}{2}$ in. Ball	Bronze or Mild Steel	2
J	Weight Arm		Mild Steel	2
K	Radius Arm		Mild Steel	2
AA	Sliding Sleeve		Brass	1
AB	Top Rocker		Mild Steel	1

Part Letter	Description	Remarks	Material Supplied	Qty.
AC	Bottom Rocker		Mild Steel	1
AD	Rocker Pivot		Mild Steel	1
AE	Rocker Pivot Collar		Mild Steel	1
AF	Rocker Standard	With Mild Steel Stud. 5BA and nut	Brass	1
AG	Throttle Link		Mild Steel	1
AH	Throttle Lever		Mild Steel	1
AJ	Throttle Link Pivot		Mild Steel	2
AK	Adjusting Link Assembly		Mild Steel	1
BA	Throttle Body		Gun Metal	1
BB	Throttle Gland Cover		Brass	1
BC	Disc		Hard Brass	1
BD	Throttle Spindle		Brass	1
BE	'O' Ring	$\frac{3}{32}$ in. cord		1
BF	Stop Valve Body		Gun Metal	1
BG	Gland Cover		Brass	1
BH	Gland Body		Brass	1
BJ	'O' Ring	$\frac{3}{32}$ in. cord		1
BK	Yoke		Mild Steel	1
CA	Valve Spindle		Rustless Steel	1
CB	Pillar		Mild Steel	2
CC	Handle		Mild Steel	1
CD	Inlet Flange	Hole to suit $\frac{1}{8}$ in. bore pipe	Brass	1
CE	Exhaust Flange	Hole to suit $\frac{5}{32}$ in. bore pipe	Brass	1
	$\frac{1}{32}$ in. Taper Pin	A. G. Thomas, Heaton Road, Bradford	Mild Steel	8
	Nut 10BA		Mild Steel	10
CC1	Nut 12BA		Brass	1
BF2	Inlet Flange Bolts 10BA	$\frac{1}{4}$ in. under head	Mild Steel	4
BF1	Stud 10BA	Standout $\frac{7}{32}$ in. ($\frac{1}{4}$ in. shortened)	Mild Steel	2
BA1	Stud 10BA	Standout $\frac{1}{8}$ in.	Mild Steel	2